Contents

List of Illustrations

Introduction — 5

1. South Wales excluding the Wye Valley — 11
2. The Forest of Dean, Wye Valley and Herefordshire — 35
3. Shropshire — 55
4. Mid and North Wales — 65
5. South Staffordshire, Worcestershire and Warwickshire — 75
6. Cheshire and North Staffordshire — 81
7. Derbyshire, Nottinghamshire and Leicestershire — 87
8. Yorkshire — 99
9. Lancashire and Cumberland — 107
10. Northumberland and Durham — 123
11. The Weald and Hampshire — 131
12. Scotland — 145

Bibliography — 153

Index — 165

List of Illustrations

Plan of Caerphilly furnace, 1764	16
Carmarthen furnace and tinworks, c. 1800	18
Plan of Hirwaun furnace, 1760	20
Coed Ithel	38
Newent	41
Guns Mill	41
Tintern furnace, c. 1800	50
Tintern today	50
Charlcotte	58
The Old Furnace at Coalbrookdale	58
Bersham	67
Dyfi	70
North Wingfield	92
Plan of Wingerworth furnace, 1758	96
Rockley	104
Backbarrow	111
Duddon	111
Leighton	115
Nibthwaite	115
Newland	118
Bonawe	146
Craleckan	148

Except where otherwise stated in the captions, the modern views were all taken by the author in 1991.

A location plan showing furnaces in each region faces the first page of each chapter.

A Gazetteer of Charcoal-fired Blast Furnaces in Great Britain in use since 1660

Philip Riden

MERTON PRIORY PRESS

Published by Merton Priory Press Ltd
7 Nant Fawr Road, Cardiff CF2 6JQ

First published by Philip Riden, 1987
Second edition, fully revised, 1993

ISBN 0 9520009 1 1

© Philip Riden 1987, 1993

Typeset at Oxford University Computing Service
Printed by Technical Print Services Ltd
Brentcliffe Avenue, Carlton Road
Nottingham NG3 7AG

Introduction

This is a fully revised, extended and illustrated second edition of a piece of work I first published in 1987. It seeks to provide an outline history of every charcoal-fired blast furnace which operated in Great Britain after 1660, to locate the site on the ground and, where appropriate, to describe what can be seen there today. It is not in any sense a detailed discussion of the evolution of the blast furnace during the seventeenth and eighteenth centuries, much less a history of even one branch of the charcoal iron industry. I originally assembled the notes as part of a long-standing interest in the compilation of output statistics for the iron industry, which, for the period covered here (chiefly 1660–1790), involves trying to establish how many furnaces were in use at any one time and how much iron, on average, each produced (cf. Riden 1977). It occurred to me that a gazetteer of sites might, on its own, be of sufficient interest to historians, geographers and archaeologists to warrant publication, and the generally kind reception accorded the first edition has encouraged me to produce this more elaborate version.

The basic arrangement and content of the material remain the same and the main change, compared with the first edition, has been the inclusion of fuller notes concerning visible remains at each site, illustrated where appropriate with a recent photograph. For every site a four- or, where possible, six-figure National Grid reference has been given as part of the heading to the entry, followed in square brackets by the number of the Ordnance Survey 1:50,000 Landranger sheet on which the reference appears, together with the (pre-1974) county in which the site lies. These features should help readers who wish to visit some of the more interesting furnaces described here and in some cases further directions have been given as to how to find the site, where to park and where to ask permission before entering private property.

The grid references can, of course, also be used to produce either hand-drawn or computer-generated maps showing the distribution of furnaces; by combining these references with the historical notes for each site it would be possible to map the changing geography of charcoal ironsmelting over time. Work of this sort lies some way beyond what I am seeking to achieve here, which is essentially to present a body of reasonably reliable data that others can exploit in different ways, although I have provided a sketch-map

INTRODUCTION

showing the location of furnaces in each of the twelve regions into which the gazetteer is divided.

My own reason for assembling the notes published here has already been mentioned. But why count blast furnaces? In 1973 G.F. Hammersley demonstrated convincingly that at no stage during the charcoal blast furnace era — roughly 1540 to 1750 — was the iron industry in Great Britain threatened by extinction for want of fuel. Nor was it, as had long been supposed, in decline in the later seventeenth and eighteenth centuries. On the contrary, in 1750, on the eve of the widespread adoption of coke smelting, the output of charcoal-smelted iron was probably greater than ever before (Hammersley 1973). A few years later, I drew together a variety of contemporary and other material to produce a new series of output estimates for the industry in the three centuries between the general introduction of the blast furnace and the beginning of reliable official statistics (Riden 1977). For the period prior to 1790, from which date the trade compiled aggregate estimates fairly frequently until the government took over the task in 1854, any attempt to measure output must be built up 'from below', i.e. by establishing the number of blast furnaces in use and multiplying that figure by an estimate of average furnace output. Before the late eighteenth century, there is only one contemporary output estimate, that compiled in 1717, which can be shown to be incomplete (Hulme 1928–9).

The estimates produced by Hammersley for the number of furnaces in use in each decade between 1550 and 1750, which also formed the basis for my own series, were derived essentially from the information collected by H.R. Schubert in an appendix to his general history of the pre-1800 iron industry (1957, 354–92). Curiously, as Hammersley observed, although his notes on individual furnaces showed quite clearly the continuing vitality of the industry after the mid-seventeenth century, Schubert clung to the older view that the industry was in decline until rescued by the adoption of coke smelting. Hammersley attempted to revise Schubert's data by drawing together local work, as did C.K. Hyde in his own lists of eighteenth-century furnaces, charcoal and coke, compiled as part of a study of technological change in the industry (Hyde 1977). I made some tentative efforts to refine the eighteenth-century data a year later (Riden 1978), by which time I had begun to appreciate the extent of Schubert's shortcomings as the main source for such an exercise.

It was at this stage that I decided that it would be desirable to compile a new gazetteer of furnaces to replace Schubert's, which could in turn form the basis of new estimates for the number of sites in use at any particular time, both nationally and in each region. Initially, I concerned myself only with the eighteenth century, in an attempt to examine in detail the transition from charcoal to coke, but later set the starting-date back to 1660, which seems to me a fairly well defined half-way stage in the charcoal blast furnace era. I have therefore excluded the first fifty or hundred years of blast furnace production in most regions, a period when, especially to begin with, the leading ironmaking district was the Weald, which has been the subject of a thorough recent study (Cleere and Crossley 1985). Elsewhere, rather less has been written of the pre-1660 industry; many furnaces are known only from isolated references in documents widely separated in date, and it is possible that new sites have still to be identified. In this situation, it would be difficult to produce a reliable national survey and pointless simply to regurgitate what others have done for the Weald.

By the mid-seventeenth century the Weald was increasingly overshadowed by other parts of the country, where furnaces become both more numerous and better documented. Whereas before 1660 continuous series of production accounts are rare for the iron industry, during the hundred years after that date several massive collections of business records not only provide a detailed picture of the economics of ironmaking but also much information about particular furnaces. This material has gradually been explored since the pioneer studies of Raistrick and Allen (Raistrick 1938, Raistrick and Allen 1939) and of R.A. Lewis (1949) and B.L.C. Johnson (1950) and is now well known, although some collections still offer scope for detailed statistical analysis, as Laurence Ince (1991) has recently demonstrated. In addition, much survives among landowners' muniments, since most furnaces were leased to professional ironmasters rather than operated, as had often been the case before the Civil War, by magnate or gentry families themselves.

There is also more aggregate contemporary data than Schubert realised. Besides the list of furnaces of 1717 and those of forges of 1736 and 1750, the Boulton & Watt papers in Birmingham Central Library contain a detailed survey of the industry in 1794 which provides a wealth of detail about individual works, whether furnaces (charcoal and coke), forges or mills. This list was published in an abbreviated form by Scrivenor (1841, 359–61), in which state it was used by Schubert and dated 1790, but has never been printed in full. A second list of charcoal furnaces closed between 1750 and 1788 has not been published at all. Where these can be checked against local record material both prove very accurate, while in other cases they provide information that cannot apparently be found locally. All these lists, especially that of 1794, underlie many of the entries here.

Even in a survey confined to the second, better documented, half of the charcoal blast furnace period, it would be impractical to base gazetteer entries entirely on record evidence. Nor is it necessary, since the industry has attracted a good deal of attention from local historians whose work, on either individual sites or regions, provides a firm foundation for a new national survey. Many articles have refined dates of opening and closure or provided more detail concerning those who operated a furnace. Since most of this work is also concerned to locate sites on the ground, something which Schubert conspicuously failed to do, it is also possible to look more closely at the geography of ironmaking in this period.

Although the gazetteer entries here have not been produced in a rigidly standard form, it should be understood that they aim to provide a certain minimum of information and no more. They are confined to furnaces and omit any but passing references to forges and mills, since the ultimate aim is to produce new estimates of pig iron output. Secondly, as far as possible, sites occupied only in the period prior to 1660 have been omitted. Sometimes it is impossible to be certain whether an earlier site was re-used and, in general, any furnace that *might* have been occupied after 1660 has been included. For each site I have tried to identify the ground landlord and occupiers (normally lessees, except in a few cases where furnaces were owner-occupied); the date of closure (and of building, if the site was newly established in this period); and any other information in the Boulton & Watt lists or that of 1717.

For the handful of furnaces that remained in use beyond 1790 I have noted the information supplied by the well known lists of furnaces of 1796, 1806 and later (Riden 1977). A few sites survived to be recorded annually in the officially published *Mineral Statistics* from 1854, from which details have been taken for this study. This series is also the main source of information for an anomalous but interesting site (Warsash, Hants.; see Chapter 11), which appears to represent an attempt to revive charcoal smelting in the mid-nineteenth century in the south of England.

For the eighteenth century, when coke was gradually replacing charcoal, there are a number of sites at which both fuels appear to have been used at different times (or even concurrently), or where it is not clear which was used. Here also my policy has been to include borderline cases and so a few entries relate to what may have been exclusively coke-fired sites, although the evidence is generally less than conclusive. I have discussed elsewhere (Riden 1992c) a group of about a dozen early coke-fired furnaces which failed during the second half of the eighteenth century, some of which may also have used charcoal.

The gazetteer is divided into twelve chapters, one of which is devoted to Scotland. England and Wales have been divided into eleven 'regions', some of which respect county boundaries, others do not. Introductory notes to each chapter have been confined to a definition of the region covered and a brief historiographical comment. For each furnace I have attempted to give the main published references, using the author-date system, but not, except in a few cases, to cite archival material. The bibliographical references are listed in a single sequence at the end of the text. This is followed by a much fuller index than that to be found in the first edition, which should be of value to anyone seeking to trace the activities of a particular ironmaster, rather than the history of a furnace.

Reconstructing the history of individual furnaces is, of course, merely the first stage in attempting to calculate the total output of pig iron in Great Britain in the period covered by this survey. This task is well beyond the scope of this survey and is fraught with problems, of which revisions to furnace chronology or the discovery of previously unrecorded sites are only two. On the other hand, one has to draw a line somewhere in an exercise of this sort and I hope shortly to publish new estimates of regional and national pig iron output for the period 1660–1790, based on the detailed data presented here.

Cardiff   Philip Riden
May 1993

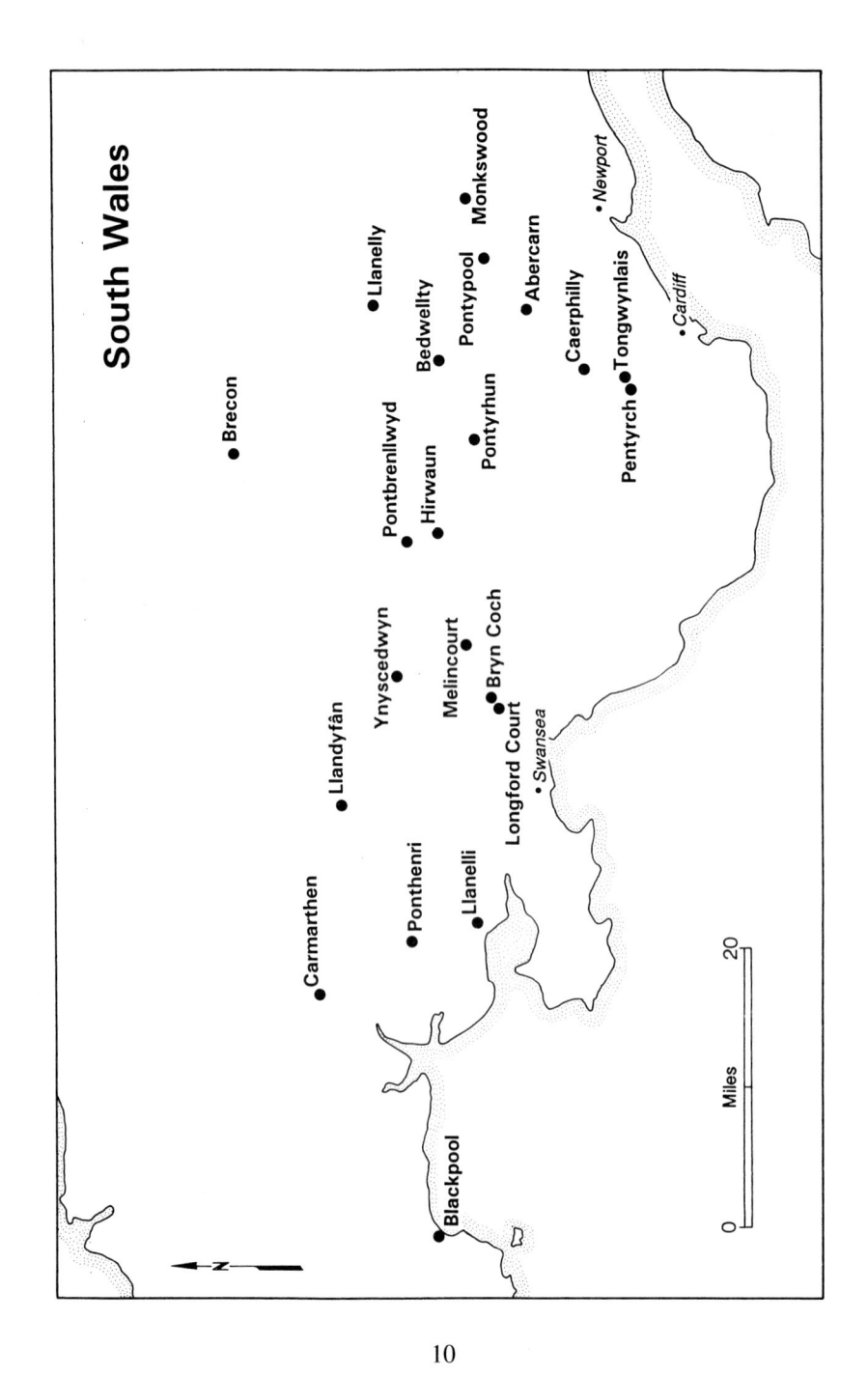

1

South Wales excluding the Wye Valley

This chapter includes all the charcoal furnaces which appear to have worked after 1660 in Carmarthenshire, Breconshire and Glamorgan, together with a possible site in Pembrokeshire, plus those in Monmouthshire apart from three in the Wye valley (Coed Ithel, Tintern and Trellech), which have been grouped with the Forest of Dean furnaces on the east bank of the Wye (Chapter 2). Charcoal ironsmelting in South Wales has tended to be neglected by local historians more interested in the better known developments after 1750 and, although there is a mass of material both in estate papers and printed sources, a modern general study is lacking. Since this gazetteer was first published in 1987 I have undertaken a good deal of work, both historical and in the field, on several of the furnaces described here, the results of which mostly remain unpublished and are unsuitable, because of the amount of detail involved, for presentation in full here. I hope in due course to make more of this work available in new accounts of individual sites.

I am indebted to Michael Evans of Carmarthen for a number of helpful comments on furnaces in the western half of the region.

Abercarn, Mon. ST 2496; ST 2195 [171]

It is now clear that there were two separate charcoal furnaces operating at different dates in the township of Abercarn in Ebbw Vale.

The earlier works was erected in 1576 by Edmund Roberts, a London merchant with a seat at Hawkhurst, Kent (Hammersley 1971, 87–8, 112; Schubert 1957, 366; Willan 1953, 119–20). Roberts died intestate and insolvent in 1579, having used the furnace before his death for casting ordnance (Hammersley 1971, 124; E.G. Jones 1939, 257–8). From 1580 until at least 1597 and possibly until his death in 1608, Richard Hanbury operated Abercarn, although it appears to have no history beyond the latter date (Schubert 1957, 366). The site of this early furnace can be located without difficulty from the survival of the name Graig Furnace on the modern map on the south bank of Nant Gwyddon about two miles NE of the village of Abercarn. The removal of soil in 1975 at the first of the map references given above located what appears to be the corner of Edmund Roberts's furnace; no report on these investigations has been published (private inf.).

A furnace at 'Abercain', 12 miles from Newport, appears in the list of closures between 1750 and 1788, when it was said to be 'down', with no details of occupier. This presumably matches references to a furnace at Abercarn in the mid-eighteenth century in local sources, which refer to an 'Abercarn Co.', which failed in 1748, and to the arrival in 1750 of a man named John Griffiths, who had apparently managed Capel Hanbury's ironworks at Pontypool (q.v.) and who built a new furnace on the east bank of the River Ebbw close to the mouth of the Gwyddon (i.e. at the modern village of Abercarn) (Pugh 1934, 61). Griffiths is said to have remained at the works for seven years; the subsequent history of the furnace has yet to be worked out, but John Darbyshire of Aston, near Birmingham, was a partner there in the 1760s (Riden 1992a, 12–13n., where the idea that John Bedford, then of Trostrey forge and later of Rogerstone, ever operated Abercarn is shown to be wrong).

In 1765–6 Joshua Glover of Birmingham was selling pig from Abercarn to the Knights' Stour Partnership forges (Ince 1991, 117) and in 1783 Joshua and Samuel Glover had a forge (but not a furnace) there, which was taking pig from Hirwaun furnace, then in the hands of Anthony Bacon (Lloyd 1906, 157–60). The 1794 survey confirms that the furnace was by then out of use, since at 'Abergwythen', 12 miles from Newport, only a forge is listed, owned by Mr Glover and occupied by Glover & Son, with four fineries, a chafery and a wire-mill. In 1801 Coxe described Abercarn as consisting of a 'pit-coal forge and charcoal wire-works', together with a disused charcoal blast furnace (Coxe 1801, I.3). The site does not appear in any of the early eighteenth-century lists.

After Bacon died in 1786 the Glovers leased Hirwaun and later bought an estate at Llwydcoed, Aberdare. In 1808, the last surviving member of the Glover family connected with the local iron industry agreed to sell the freehold of the Abercarn works to Richard Crawshay of Cyfarthfa, who subsequently made a gift of the estate to his son-in-law Benjamin Hall. The forge remained in the ownership of the Hall family throughout the nineteenth century and was operated by various lessees as a tinworks (Lloyd 1906, 157-60; Pugh 1934, 89). The area on the east bank of the Ebbw occupied by the works has now been redeveloped for new industry, destroying all trace of the early furnace or forge.

Bedwellty, Mon. SO 165043 [171]

A furnace in the parish of Bedwellty, in the Sirhowy valley, is mentioned in Exchequer depositions in 1597, when it was said to be in the hands of John Challenor, a London haberdasher, and Thomas Moore, a Bristol ironmonger (Schubert 1957, 367; Hammersley 1971, 88). It is mentioned again in Challenor's will of 1606 but not that of his son, William Challenor (1620). The approximate position of the furnace is indicated on the modern map by the name Pontgwaithyrhaiarn (i.e. Ironworks Bridge) for a crossing of the Sirhowy about three miles south of Tredegar, where indeterminate earthworks can be seen today on the east bank of the river.

The site would not merit inclusion here were it not for a confused statement in an essay written for a local eisteddfod in 1862 (Morris 1868, 20–4; cf. Jones 1969, 110–12), based on oral evidence collected about 1825–30, that the furnace at Pontgwaithyrhaiarn was in use in the 1740s, although for how long is not clear; remains of the structure were apparently also visible in the 1820s. A later history of Tredegar (Powell 1885, 19–22) complicates the issue further by claiming that the site was in use in the late seventeenth century, while a modern antiqury notes references to an 'Old Iron Works' in Bedwellty in a rental of 1727 (Jones 1969, 26). All this requires confirmation from record evidence but it appears that the furnace may have operated, if only for short periods, after 1660.

Blackpool, Pembs. SN 145060 [158]

M.C.S. Evans (pers. comm.) suggests that there may have been a furnace at Blackpool, near the site of the forge near Canaston Bridge, which appears in early eighteenth-century lists of forges. The only evidence he has so far discovered is the name 'Furnace Farm' in eighteenth-century deeds and on the first edition of the 1in. Ordnance Survey map (*c.* 1830). There is no suggestion in any of the nationally compiled surveys that smelting took place here, although Rhys Jenkins evidently believed that there was also a furnace at Blackpool (Hulme 1928–9, 35).

Brecon, Brecs. SO 052298 [160]

The furnace at Brecon was established by a partnership between Benjamin Tanner, ironmonger of Brecon, and Richard Wellington of Hay (Brecs.), who in 1720 took a ninety-nine-year lease of the site of a tucking-mill at Tŷ Watkin, on the River Honddu north of the town, the lessor being Richard Jeffreys of Brecon Priory. The furnace operated in conjunction with a forge built by the same partners at Maes y Wern Isa at Pipton, on the east bank of the River Llynfi, on land belonging to Henry Williams. The Tanner family continued to operate both works until Benjamin's son William, with the consent of his partners, sold the furnace and forge in 1753 to Thomas Daniel and Richard Reynolds sen., both Bristol iron merchants, who almost immediately disposed of them to Thomas Maybery. He in turn transferred them in 1755 to his son John Maybery, who two years later built a new furnace at Hirwaun (q.v.). In 1760 Maybery took his brother-in-law John Wilkins into partnership and, at about the same time, converted the Brecon site into a forge refining pig from Hirwaun, Pipton forge being abandoned (Rees 1968, 308; Schubert 1957, 368; Minchinton 1961, 9–10).

This account corresponds with the evidence of the 1794 list and that of furnaces closed between 1750 and 1788; the furnace is too late to appear in the 1717 list. In 1794 a forge alone is listed at Brecon, owned by Lord Camden (the successor of the Jeffreys family), occupied by Wilkins & Jeffries, with a finery as the only plant. The works was located a mile from Brecon, which can only refer to the site on the Honddu, where the name 'The Forge' appears on modern maps, and not the Pipton site. In the list of closures between 1750 and 1788 the entry reads 'Brecknock furnace 1760 ... Forge'. William Rees, however, suggested (1968, 308), without citing any authority, that David Tanner took over the furnace at Brecon and forge at Pipton sometime after 1770, with both works finally shutting in 1799 after Tanner was declared insolvent. There appears to be no record evidence for this, and the name 'Tanner's Forge' for the site at Brecon, which is that used in 1794, simply refers to the original builder in 1720.

At the site today the main property is still known as Forge Farm, which may have been built as a manager's house. There are also some cottages nearby which are probably contemporary with the furnace, while the junction at which the lane down to the hamlet leaves the main A470 is called Furnace Gate and is the site of a toll-bar. There are no obvious surface remains of the furnace itself, which is regrettable, since in John Lloyd's time it was 'comparatively uninjured by the ravages of time and the hand of man, present[ing] to view to-day its original appearance, and is well worth a visit' (Lloyd 1906, 1). He was able to publish a photograph of the structure, then apparently almost intact, showing tapping- and blowing-arches with cast-iron lintels bearing the date 1720 (which one suspects was touched-in for the occasion). Nor is there any trace today of the leat or mill-pond to which Lloyd referred.

Bryn Coch, Glam. SS 7499 [170]

Phillips (1925, 286-8) appears to be the first writer to note the existence of a furnace near Bryn Coch Farm in the lower Neath Valley, which fails to appear in any of the nationally compiled lists but whose site (where there are no obvious surface remains) is recalled by the name 'Old Furnace' on the six-inch OS map of *c.* 1880. The only definite reference to this furnace appears to be a lease of the furnace in 1772 from Phillip Williams of Dyffryn Clydach (Glam.) to Coles, Lewis & Co. for 35 years; the same firm also had Melincourt furnace (q.v.) higher up the valley and several forges and tinplate works in the area. Phillips suggests that the furnace might have been in existence in 1735 or earlier, given what is known of the families concerned; this was not sufficient warrant for A.H. John (1980, 26) to say that Bryn Coch furnace was built that year. The site was presumably abandoned fairly shortly after the lease of 1772, although it does not appear in the list of closures between 1750 and 1788. It is quite distinct from the later coke furnaces at Neath Abbey (Ince 1984) and apparently also from a seventeenth-century furnace near Longford Court (q.v.).

Caerphilly, Glam. ST 142878 [171]

The furnace here was built in 1680 and operated throughout the eighteenth century in tandem with forges at Machen and Tredegar Park (Schubert 1957, 369; Williams 1959-60; Rees 1968, 312-17). In the 1690s the works were operated by a partnership that included John Morgan of Tredegar Park, the owner of the furnace and forges, whose son Sir William Morgan leased the works in 1732 to James Pratt. On the expiry of Pratt's lease in 1747 the works were assigned to Thomas Morgan of Ruperra (son of Sir William), Hugh Jones of Gelliwastad, and Samuel Pratt. The partnership traded as James Pratt & Co. until the death of Samuel Pratt, when it was reconstituted as Hugh Jones & Co., continuing as such until 1764, when Hugh Jones left. It is presumably this firm, described as Hugh Jones & Co. of Machen, which appears in the Knights' accounts between 1749 and 1754, supplying pig from Caerphilly to the Stour Partnership forges (Ince 1991, 117).

SOUTH WALES

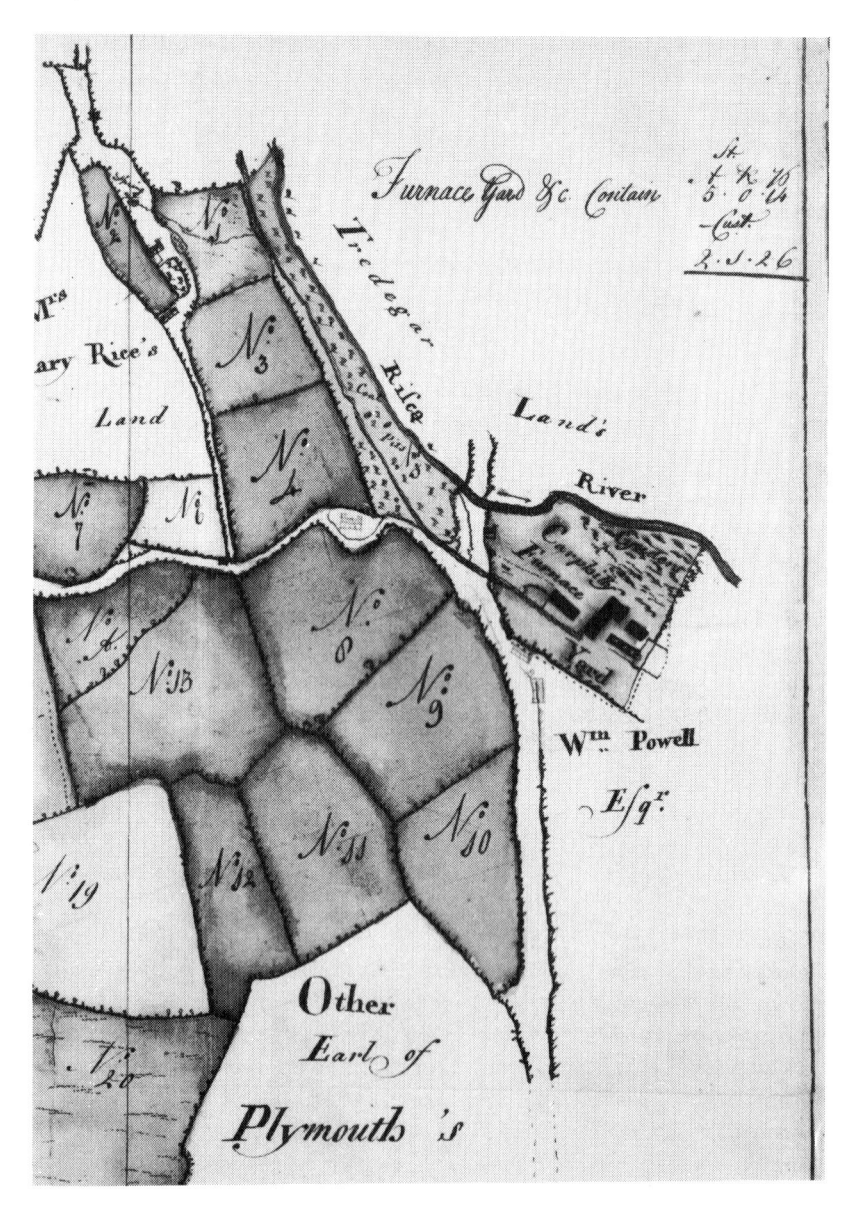

Plan of Caerphilly furnace, 1764. A detail from a Tredegar Estate plan showing the furnace and ancillary buildings on Nant yr Aber near Hendredenny. The road passing the furnace to the west is the present B4263. *(National Library of Wales)*

In 1764 Morgan leased the works to John Maybery of Brecon for 21 years, a period later extended to 42, and entered into a new agreement in 1775 concerning Caerphilly furnace. By 1789, however, the then head of the Tredegar Park family, John Morgan, leased the entire group of works to the Bristol ironmasters James Harford, Philip Croker and Truman Harford for 21 years. These details correspond with the evidence of the 1794 list, in which Caerphilly furnace is said to be owned by John Morgan and occupied by Harford & Partridge, the plant consisting of an 'old' charcoal furnace and a rolling- and slitting-mill. In 1796 Caerphilly was said to produce 600 tons a year, which may be an actual output figure but is probably extrapolated from a weekly figure of 12 tons, since in 1717 the furnace was said to produce only 200 tons a year. Smelting appears to have ceased at Caerphilly in about 1819 (Rhys 1858; Richards 1968).

The furnace is shown in some detail on a Tredegar estate plan of 1764, on which the furnace, blowing-house with charging-bridge over, casting-house, cinder bank, storehouse and calcining kilns can be made out, as well as a leat coming down the valley from the north-west. The works stood on the south bank of Nant yr Aber immediately downstream from the bridge which carries the B4263 over the brook at Trecenydd, west of Caerphilly (cf. Rees 1969, 60). The site, which is privately owned but accessible, remains unbuilt upon. In addition to a large quantity of slag and masonry and firebrick rubble, it is also possible to locate what appears to be the base of the furnace itself, immediately in front of a bank from which it would have been charged.

Carmarthen, Carms.   SN 420206 [159]

Carmarthen furnace was established by Robert Morgan in 1747, when he took a lease of the former Priory Mill about half a mile north-east of the town, where he built a furnace, two rolling-mills and other works. Morgan, variously described as a 'merchant' or 'gentleman' of Carmarthen, was the second son by his first marriage of Christopher Morgan, 'merchant' or 'shopkeeper' of Kidwelly, where Robert was born in 1708. In 1760 he enlarged his activities by taking a lease of Blackpool forge (Pembs., q.v.) and between 1759 and 1761 built a tinplate works at Carmarthen. He also operated forges at Kidwelly, Whitland and Cwmdwyfran. In 1763–7 Morgan was selling pig from Carmarthen to the Knights' Stour Partnership forges, having for the previous twenty years supplied them from the furnace at Ponthenri (q.v.), near Kidwelly (Ince 1991, 117).

Carmarthen furnace and tinworks, c. 1800. A watercolour view looking south-west along the leat leading to the furnace water-wheel. *(National Library of Wales)*

Robert Morgan's will is dated 1777 and was proved the following year, leaving the business to his son John, who continued it on the same lines. A trade token issued by him shows the furnace at Carmarthen. John Morgan died in 1805, leaving the business to various relatives, but the works foundered and the estate was the subject of Chancery proceedings. About 1824 the firm then operating the tinplate works at Carmarthen built new works at Aberavon (Glam.) and transferred plant and employees there. The furnace and tinworks apparently remained in use until 1821, the former still using charcoal as fuel (Green 1915; James 1976; L.J. Williams 1960–1a and b; L.W. Evans 1938; M.C.S. Evans 1967; Rees 1968, 309–12; Lloyd 1939, 323–33).

The 1794 list locates the works half a mile from Carmarthen, owned and occupied by Mr Morgan, with a single charcoal furnace, described as water-powered and 'old'. There were two fineries, a chafery and tin-mill at the forge. The site, which remained in use as a tinplate works until the late nineteenth century, is now occupied by a branch of Jewson's, the builder's merchants, with whose permission the substantially complete remains of the furnace may be inspected behind their warehouse. At the level of the road which runs above the site, a furniture showroom was built on top of the furnace by Jewson's predecessors. Across the road, some of the buildings directly opposite the showroom may incorporate remains of a charging-house or store-rooms from the furnace (cf. James 1980, 56).

Hirwaun, Brecs. SN 958058 [160]

The furnace at Hirwaun, on the north bank of the River Cynon in the Breconshire parish of Penderyn (rather than south of the river in Aberdare, Glam., where the modern village of Hirwaun grew up) was established in 1757 by John Maybery on land leased from Lord Windsor. In 1760 he took Mary Maybery and his son-in-law John Wilkins into partnership. Four years later he leased the forge on the Morgan estate at Tredegar Park and in 1777, again in partnership with Wilkins, leased the forge at Machen (cf. Caerphilly). After this date Hirwaun seems to have become unprofitable and in 1778 was taken over by Walter and Jeffrey Wilkins, the Brecon bankers. In 1780 the lease was assigned to Anthony Bacon, who apparently revived the works and operated them up to his death in 1786, when his sons, who were minors, leased them to Samuel Glover of Abercarn (Minchinton 1961, 10–11, 15). It is Glover who appears as occupier at Hirwaun in 1794, with

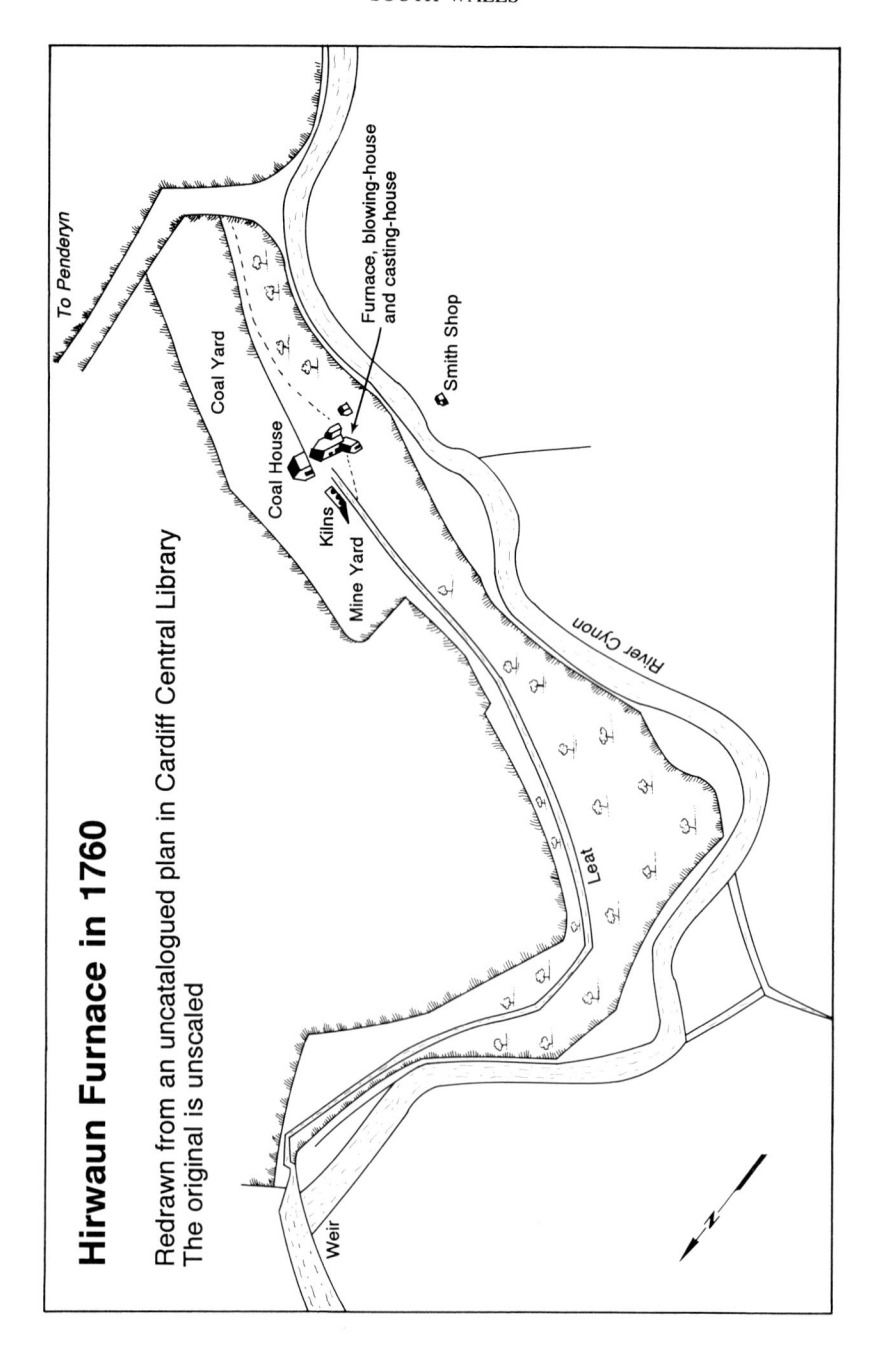

SOUTH WALES

Hirwaun Furnace in 1760

Redrawn from an uncatalogued plan in Cardiff Central Library
The original is unscaled

Lord Cardiff as owner, the plant consisting of a coke furnace blown by steam, said to have been built in 1758.

Hirwaun has normally been coupled with Dowlais as one of the two earliest coke furnaces in South Wales (e.g. Riden 1978), but the evidence that coke was used from the start is not completely clear. The near-contemporary and generally reliable Breconshire historian Theophilus Jones (1805–9, 640–1) states quite clearly that Hirwaun 'was used for the purpose of smelting ore with charcoal for many years, and afterwards with pit or mineral coal, being supplied with blast by means of a water wheel'. Minchinton suggested that it may have been Anthony Bacon who converted the works to coke-smelting after 1780. Recently, however, new evidence has come to light showing the Hirwaun was using coke in the mid-1760s (ex inf. Chris Evans from the unpublished diary of Charles Wood in the Glamorgan Record Office), which may be regarded as sufficient to discount Jones's comment. Alternatively, it is possible that both charcoal and coke were used in the early years until Maybery was confident of the success of the new process.

Hirwaun remained in use during the Napoleonic Wars but was a victim of the slump that followed the coming of peace in 1815. Two years later the works was acquired by William Crawshay of Cyfarthfa, who rebuilt the site in the early 1820s with a blowing-engine and four new furnaces. The works closed in 1859, following the expiry of the lease of 1758, although there were sporadic attempts at revival until the end of the century.

The remains on the site today (which form a scheduled monument), although impressive, appear to belong entirely to the works built by Crawshay in the 1820s. The only clue to the original layout of the works comes from a plan of 1760 (Cardiff Central Library, Local Studies Dept, no reference number, cannot at present be found; I am indebted to Howard Llewellyn for bringing this item to my attention). This marks (in an ill-informed fashion) a furnace with adjoining blowing- and casting-houses, a leat from the Cynon (whose course here has been altered since 1760), a bank of calcining kilns, yards for ironstone and 'Coal', and a 'Coal House'. The latter probably points to the use of charcoal, since coke seems usually to have been stored outside, rather than under cover.

Since Hirwaun and Dowlais are normally linked together in accounts of the later South Wales iron industry, the evidence for coke-smelting from the start at the latter furnace (SO 0608) should perhaps be set out here. Dowlais was blown-in in 1760, having been built by a company headed by Thomas Lewis of New House, Llanishen, near Cardiff, who also had Pentyrch furnace (q.v.) and New Forge, Cardiff (John 1980, 27–8), and was associated with the west Glamorgan firm of Coles, Lewis & Co. The partnership articles were executed in 1759, although the main lease under which the partners got minerals was granted as early as 1748 by Lord Windsor to Thomas Morgan of Ruperra, a member of the Tredegar Park family, who was then operating Caerphilly furnace (q.v.), but Morgan had made no use of the concession. A new mineral lease of 1763 includes a plan of Dowlais furnace (Elsas 1960, frontispiece; E. Jones 1987, 4), while another schedule of the premises was drawn up three or four years later. The latter includes a coking yard, three calcining kilns, a moulding room and casting yard, all of which point to coke-smelting, as conversely does the absence of a charcoal store (England 1959, 41–4; cf. E. Jones 1987, ch. 1).

Kidwelly: see Ponthenri

Llandyfân, Carms. SN 658170 [159]

Evans (1973, 146) draws attention to two letters of 1756 and 1758 which refer to disused furnace sites at Llandyfân, where two forges can also be documented from the late seventeenth century to the early nineteenth. The details are not sufficient to locate or date the sites exactly, but there are appears to have been one, or possibly two, furnaces in this area prior to the mid-eighteenth century.

Llanelli, Carms. SN 504015 [159]

Symons (1979, 96n.) suggests that the coke furnace at Cwmddyche on the outskirts of Llanelli, generally assumed to have been established on a new site in 1793 by Alexander Raby and two associates, may have been preceded by a charcoal furnace in approximately the same place. A letter of 1762 from Daniel Shewen to his brother Joseph refers to 'a bad time ... the furnace being idle for 2 years', apparently meaning a furnace on the Mansel estate in Llanelli, although this is not at present entirely clear (ex inf. M.C.S. Evans).

Substantial remains of Raby's furnace can still be seen in the valley below the B4309 in the hamlet of Furnace on the NW side of Llanelli, together with a storage pond and dam associated with the works (Hughes and Reynolds 1988, 19).

Llanelly, Brecs. SO 232138 [161]

Schubert (1957, 371) relates a local tradition that there was a furnace at Llanelly as early as 1590, which was rebuilt on a new site in 1606. The earliest record evidence appears to be contained in articles of agreement of 1663 under which Capel Hanbury was to have liberty to build a weir to divert water from the River Clydach to his ironworks at Llanelly (Gwent RO, D8A, JCH 1272); the furnace is also mentioned in 1684 and again in John Hanbury's account of his ironworks of 1704–8 (van Laun 1989, 25–6). Llanelly was listed in 1717, with an output of 400 tons, and in 1794 as an 'old' charcoal furnace, owned by the Hanbury family, occupied by David Tanner, with two melting furnaces but no other forge plant. In the 1720s John Hanbury was supplying pig from Llanelly to the Foleys' forge at Llancillo (Johnson 1953, 142) and in 1745–7 to the Knights' Stour Partnership forges (Ince 1991, 117).

Although the charcoal furnace is listed in 1794 and shown on an estate map of about the same date (van Laun 1989, 25), it may have been abandoned some years before, since a furnace at 'Llanelly' appears in the list of closures between 1750 and 1788 with the comment 'down'. Alternatively, this could refer to a charcoal furnace at Llanelli, Carms. (q.v.), which may have preceded Alexander Raby's coke furnace of 1793. If the Breconshire furnace was still in use in the mid-1790s, it must have closed shortly afterwards, since a coke furnace was built about half a mile higher up the Clydach valley in 1793 (Wilson 1989) and it seems unlikely that the earlier site would have been kept in use for long, particularly as Tanner became bankrupt in 1798 and Hanbury would have had to find a new tenant. Pig from both furnaces was refined at a forge a short distance down the valley from the charcoal furnace.

The site of the earlier furnace in the Clydach Gorge can be located without difficulty from the estate plan of *c.* 1795 (van Laun 1989, 25), about a hundred yards SW of Clydach House, a fine seventeenth-century house occupied by the clerk of the works. Over the main doorway is a stone plaque bearing the initials F.L. (Francis Lewis), the Lewis arms and the date 1693. The furnace stood immediately in front of a stone-faced bank, much of which can be traced behind disused industrial premises known as the Sale Yard, to the rear of a row of modern houses. The furnace itself stood in the garden of one of these properties ('Mandalay') and there appear to be no surface remains, although the truncated abutment of the bridge linking the bank and the furnace can be seen. At the back of the bank, the plan of *c.* 1795 shows a charcoal store, which must have stood roughly where an abandoned portable building has been dumped, and behind that stands a substantial bank of limekilns, fed from higher ground to the north.

The substantial remains of Clydach Ironworks, the coke-fired furnaces higher up the valley (SO 229132), have recently been excavated and consolidated for public display by Blaenau Gwent Borough Council (Wilson 1989); both this site and the earlier furnace are scheduled monuments. Of the forge lower down the valley (SO 236140) little remains apart from Forge House, which was once a tinworks, and the retaining wall of the pond. All the considerable number of features of interest in the Clydach Gorge are well described and interpreted in an excellent guide published by the Brecon Beacons National Park (van Laun 1989).

Longford Court, Glam. SS 7398 [170]

Several writers have noticed a lease of 1694 (Nat. Lib. Wales, Coleman 829; cf. Phillips 1925, 285–6; Rees 1974, 149–50; W. Rees 1968, 304) of a parcel of land 'whereon an iron melting furnace formerly stood' for the erection of a copper mill. The land abutted the River Clydach east, the highway from Neath to 'Courtride' (i.e. Cwrtrhydhir, Longford Court) west, lands and houses south, and a brook leading from the highway to the Clydach and a garden north. This appears to point to the existence at some date earlier in the seventeenth century of a furnace which at present has no recorded history. The site is not identical with either that at Bryn Coch (q.v.) or (it seems, but cf. *Trans. Neath Antiquarian Soc.*, 2nd series, 3 (1932–3), 82–3) the later coke furnace at Neath Abbey (Ince 1984).

Melincourt, Glam. SN 824018 [170]

A furnace near the confluence of Melincourt Brook and the River Neath, near the modern village of Resolven, was blown-in in 1708 by John Hanbury of Pontypool (Schubert 1957, 382). The furnace was listed in 1717 with an output of 200 tons. In 1794 it appears as Melin y Court, with the owner as Lord Talbot (in error for Lord Vernon) and the occupier as 'Dr Lettsom' (i.e. John Coakley Lettsom (1744–1815), Quaker physician and son-in-law of the previous lessee, John Miers; cf. Raistrick 1968, 304–5). The furnace was then using coke but was still blown with water and was described as 'old'. There was no forge. In 1796 output was listed as 648 tons, which is probably a true annual figure, rather than a multiple of weekly production, and possibly points to conversion to coke.

Phillips (1925, 294–9) established that Melincourt was built by Sir Thomas Mansel of Margam, 1st Lord Mansel, and was later worked by Mansel himself in conjunction with a forge at Aberavon. Accounts survive among the Margam muniments for the furnace in 1718. In 1736 the furnace was leased to Thomas Popkin, and in 1747 to Rowland Pytt of Gloucester, who was joined in 1748–9 by Thomas Lewis. John Miers, a London merchant and pioneer of tinplate manufacture in west Glamorgan, joined the company between 1754 and 1758, as did William Coles, the firm thereafter being known as Coles, Lewis & Co. In 1766 a new lease was granted to John Miers, William Coles and Thomas Lewis and the firm was subsequently known as John Miers & Co. According to a contemporary traveller, Melincourt was rebuilt as a coke furnace in 1788 and continued in blast until 1808; in fact it seems more likely that conversion took place after a new lease was made to John Miers's executors in 1793 (Green 1980–1, 70-5).

At the site today (cf. Green 1980–1; Hughes and Reynolds 1988, 18) there are substantial remains of the furnace itself and other buildings, although considerable clearance, if not excavation, would be needed before these could be fully interpreted. The site, a scheduled monument, is readily accessible from the minor road which runs to the north of Melincourt Brook from the B4434 between Resolven and Clyne.

Monkswood, Mon. *c.* ST 3302 [171]

A furnace and forge at Monkswood, west of Usk, were built in 1564 and were thus amongst the first works of their kind in South Wales (Hammersley 1971, 110). Most of what is known of them during the rest of the sixteenth century comes from a series of Exchequer actions involving the Company of Mineral & Battery Works and their members, including Richard Hanbury, the papers from which have been used by several writers (Bradney 1904–33, I.432–4; Schubert 1957, 382; Donald 1961; Rees 1968; Hammersley 1971). After various changes in ownership the furnace at Monkswood, plus two associated forges, were in Hanbury's possession in 1597 and probably remained so until his death in 1608 (Schubert 1957, 382).

The later history of the works has never been established: Bradney (1904–33, III.72) merely suggested that they were probably removed to Pontypool, since this was for many years afterwards the main centre of the iron industry in the area. On the other hand, Monkswood appears in the list of furnaces closed between 1750 and 1788 but with no occupier's name. Further evidence suggesting either the survival or, perhaps more likely, revival of Monkswood comes from a survey of the Duke of Beaufort's estate in the neighbouring hamlet of Glascoed of 1772–9 (National Library of Wales, Badminton Maps 10, examined by kind permission of His Grace the Duke of Beaufort). Two plans in this volume describe two different lanes leading from Glascoed to Monkswood as 'The Road to the Old Furnace'. It is just possible that the site of a furnace which had been out of use for perhaps a century and a half would still be a landmark on a plan of this date but it seems more likely that the works had in fact been in use at some date in the eighteenth century. It was not, however, included in the list of 1717.

The two lanes in Glascoed said to lead to the old furnace converge to cross Berthin Brook at SO 334027, from where a track leads up to the A472 and continues further north as a minor metalled road known as Rumble Street. Courtney (1991, 69) was unable to locate any field evidence for a furnace in this area, but it may be worth drawing attention to a small, triangular enclosure of ground alongside the track which runs between Berthin Brook and the A472, which appears once to have had a building on it and whose position fits the evidence of the estate plan. A furnace here could have been powered by the small stream which still flows alongside the track down from the main road to Berthin Brook. An alternative site has been suggested by A.G. Mein (*Archaeology in Wales*, 22 (1982), 47–8) at the more northerly of the two bungalows which stand on the lane a few hundred yards east of Rumble Street (SO 336027), where he found slag and possibly rubble. It is difficult to identify a source of water-power for a furnace on this site, which would not have lain directly on the route identified on the estate plan as the 'Road to the Old Furnace'. In 1880 the Ordnance Survey named a building on the site of the present bungalows as 'Bailey Coalhouse', although the significance of this is not clear.

Pentyrch, Glam. ST 122832 [171]

A furnace at Pentyrch, on the right bank of the Taff about six miles north of Cardiff, appears to have been established *c.* 1740 by Thomas Lewis of New House, Llanishen, and Thomas Price of Watford, Caerphilly; it is referred to as the 'New Furnace in the parish of Pentyrch' in 1745 and marked as 'Pentrich Forge' on Yates's map of 1799 (Chappell 1940, 23–4). The site was leased from William, Lord Talbot. In 1775 the works were held by Nicholas Price jun. and in 1796 apparently came into the hands of

William Lewis, son of Wyndham Lewis, third son of Thomas Lewis, probably by his marriage to Mary Price. In 1805 the works were sold to Harford, Partridge & Co., who were then operating the tinplate works about a mile down the valley at Melingriffith (Rees 1968, 300).

In the 1794 list the furnace was said to be owned by Lord Talbot and occupied by Mr (or possibly Messrs) Lewis, but the only reference to plant is a figure 0 in the column for coke furnaces and the additional comments 'water' (blown) and 'old'. There appears to be no evidence that coke was being used at this date at Pentyrch, which was probably then a charcoal furnace out of blast, as it certainly was when the 1796 survey was compiled. Chappell (1940, 25–9) made use of an account book from Pentyrch covering part of the 1790s (in private hands), which apparently referred to both a melting furnace and the introduction of puddling there in 1792, but only occasionally mentioned smelting. He believed that most of the pig used at Pentyrch was imported from Lancashire or Bristol, and that most of the bar was sent to Melingriffith. After 1805 both Pentyrch and Melingriffith were enlarged and, according to Chappell, smelting increased, using both charcoal and coke; a tour account of 1802 indicates that charcoal alone was then in use (ex inf. Peter Wakelin). Pentyrch does not appear at all in the 1806 list. It was, on the other hand, included in all the surveys from 1823 onwards and in *Mineral Statistics* down to the abandonment of the last furnace in 1885 (cf. Chappell 1940, 55–70).

The site of the ironworks was for a long time assumed to lie beneath the Heol Berry housing estate, built by Cardiff Rural District Council shortly after the Second World War. In 1989, however, contractors erecting the houses which now stand between Heol Berry and the Taff found substantial remains of two of the three nineteenth-century furnaces from Pentyrch, which could be related to features shown on old photographs and the first edition of the 1:2500 Ordnance Survey plan. Although some further investigations were conducted by the Glamorgan-Gwent Archaeological Trust, no definite remains of the eighteenth-century furnace were located. Nor was anything found (notwithstanding the claims in Egan 1990, 205, where the grid reference is also substantially inaccurate; cf. *Archaeology in Wales*, 29 (1989), 70, for a fuller account with less dogmatic conclusions) to confirm the supposition (which in itself seems reasonable) that the works established in 1740 was on the same site as a furnace in the parish of Pentyrch in use between the early 1560s and 1616 (Riden 1992b).

Pontbrenllwyd, Brecs.   SN 9508 [160]

References to a blast furnace at Pontbrenllwyd, in the parish of Penderyn (which is not mentioned in Theophilus Jones's generally thorough county history of 1805–9), appear to originate in the earliest historical account of Aberdare, lower down the Cynon valley, written for a local eisteddfod in 1853, where it is said that a furnace was built 'at Hirwaun' (which lies between Aberdare and Pontbrenllwyd) in 1666 by a gentleman named Maybery. The fuel used was charcoal, ironstone was brought to the furnace on horseback, the blast 'was provided by two men with smith's bellows', and the output of iron was a ton a week (Llewelyn 1853, 46). At first sight, given the references to Hirwaun and Maybery, this appears to be a garbled account of the establishment of Hirwaun furnace (q.v.) nearly a century later. The story reappears in a Welsh history of Glamorgan of 1875 (which draws heavily on early eisteddfod essays for the history of local ironworks), where the author adds the comment that the iron made at this furnace was taken to Brecon (D.W. Jones 1875, 204). This again suggests confusion with Hirwaun, from where pig was also taken to a forge at Brecon on the site of the former furnace there (q.v.). Jones, however, clearly dates the building of Maybery's earlier furnace to 1666 and distinguishes it from the later Hirwaun site.

The most detailed account appears in a Welsh history of Penderyn (D. Davies 1905, 42), which identifies the builder of the furnace as the great uncle of the Rev. Charles Maybery (rector of Penderyn from 1832 until his death in 1871 (Ibid., 100)) and locates the site at Pontbrenllwyd, near Llygad Cynon. Both names can be found on the modern map and are over a mile upstream from Hirwaun furnace. Jones could not offer a date for the abandonment of the Pontbrenllwyd site but suggested that in 1757 the works were removed to Brecon, which was more convenient for the supply of wood. This statement is clearly confused but Jones appears to be linking the closure with the establishment of Hirwaun furnace, which he describes next.

No record evidence has yet been located to confirm this well established local tradition that the Maybery family were making iron in the area nearly a century before they acquired the Brecon and Pipton works and then built the furnace at Hirwaun. Nor is there any obvious evidence on the ground for a furnace near the bridge over the Cynon at Pontbrenllwyd.

Ponthenri, Carms. SN 478086; SN 474092 [159]

Although a Welsh poem attributes the establishment of furnace at Ponthenri to a Swedish immigrant during the reign of Elizabeth, there is no firm evidence for its operation until 1611, when it was in the hands of Hugh Grundy of Llangendeirne, who in 1620 was granted a patent for 'charking earth fuel' for use in ironsmelting. Difficulties with other local landowners led to its closure in 1629 and the history of the site in the following years is obscure (M.C.S. Evans 1967, 30–9; the entry in Schubert 1957, 379, under Kidwelly is confused and mostly relates to the earlier works at Whitland Abbey: Evans, 22–6). The property descended to Lucy, the granddaughter of Hugh Grundy, who married Anthony Morgan, but it is not clear whether the Morgan family actually worked the furnace.

In 1696 Thomas Morgan and his mother Elizabeth, widow of John Morgan, leased the property to Thomas Chetle, a descendant of a Worcestershire family living at Walhouse in the parish of Hanbury. The lease of 1696 is the family's first known connection with the Carmarthenshire iron industry. A new furnace was built at Ponthenri and for a time Zachary Downing of Halesowen (Shropshire) was in partnership with Chetle. Both Downing and Chetle's son Peter, who was born in 1681, were at different times in charge of the furnace. In 1710 Thomas Chetle surrendered his lease of Ponthenri and a new grant was made to his son.

The furnace is probably that listed as 'Kidwelly' in 1717, then said to be producing 100 tons annually; Chetle had a forge at Kidwelly (about four miles from Ponthenri) but there appears never to have been a furnace there. Chetle's other forge was at Llandyfân. In 1729 Peter Chetle sold all his ironmaking interests in Carmarthenshire, including Whitland and Kidwelly forges and Ponthenri furnace, to Lewis Hughes of Carmarthen; he died the same year and was buried at Carmarthen.

The various works grouped into a single business by Chetle later passed, precisely when does not seem to have been established, to Robert Morgan, who in 1743–58 and again in 1761–3 was selling pig to the Knights from a furnace named in the Stour Partnership accounts as 'Kidwelly' (Ince 1991, 117), which again probably refers to Ponthenri, from where iron would have been taken to Kidwelly for shipping up the Severn. In 1747 Morgan took a lease of the priory mill at Carmarthen (q.v.), where he built a new furnace, which in turn appears in the Knights' account from 1763 onwards. Ponthenri may therefore have been abandoned about this date, which is somewhat later than that deduced by Evans; it also seems that, for a time, Morgan operated two furnaces concurrently in Carmarthenshire. The appearance of a furnace at 'Kidwelly' in the list of closures between 1750 and 1788 presumably once again refers to Ponthenri.

Evans (1967, 30, and pers. comm.) located two furnace sites at SN 478086 (Furnace Farm) on the left bank of the Gwendraeth Fawr and SN 474092 (Hen Ffwrnes, i.e. Old Furnace) on the opposite bank (cf. Roberts 1979–80 and D. Thomas 1905, 10-13). It is not entirely clear which of these sites is the earlier and which the new furnace built under the lease of 1696.

Pontypool, Mon. SO 268003 [171]

There appear to be three separate charcoal furnace sites near the modern town of Pontypool, two of which, at Cwm Ffrwd-oer (SO 2601) and Pontymoel (SO 2900), were operated by Richard Hanbury and his associates in the later sixteenth century and have no history after his death in 1608, although there was a forge at Pontymoel until 1820 (Schubert 1957, 373, 384; Hammersley 1971, 112).

The third site, which in his summary tables Schubert said worked between 1576 and 1831 but which was omitted from his gazetteer (1957, 358–9, 384), was at Trosnant, about a mile west of the town, where the name Old Furnace survives on the modern map. This was apparently also established by the Hanbury family, by whom it was still owned in 1794, although exactly when is not clear. It may be one of the sites referred to in late sixteenth-century sources, or it may be a replacement for an earlier furnace elsewhere in the Pontypool area. In 1717 a furnace at Pontypool (clearly by this date the Trosnant site) was said to be producing 400 tons annually and the forge, presumably at Pontymoel, some 350 tons of bar iron. Between 1737 and 1757 John Hanbury sold pig from Pontypool to the Knights' Stour Partnership forges (Ince 1991, 118).

In 1794 the single charcoal furnace at Trosnant was described as 'old', and the forge plant consisted of three fineries, two chaferies, a wire-mill and a rolling- and slitting-mill, with the note 'Tin Mill' written in the margin. (Pontypool was the scene of the Hanburys' early adoption of tinplate manufacture at the beginning of the eighteenth century.) The works were then operated by David Tanner. The exact date of closure is not clear; no evidence has been found to support Schubert's suggestion of 1831. Tucker and Wakelin (1981) print a tourist's account of the works in 1782.

D.M. Rees (1975, 268) located part of the furnace, charging-platform and other features at Trosnant in 1973 but since then there has been extensive tipping on the site and little can now be made out. The only available record of what was visible twenty years ago seems to be a line drawing made in the early 1970s by a local artist (Nichols 1978, title-page), which confirms Rees's description of the remains. The furnace was powered by water stored in two large ponds higher up Cwm Glyn, which are now drained, although remnants of the dam walls are still visible.

Pontyrhun, Glam. SO 065028 [170]

The list of closures between 1750 and 1788 includes an entry under South Wales which reads 'Jenkins furnace now Plym ... Furnace', which can apparently best be interpreted to refer to a charcoal furnace near the later Plymouth Ironworks on the Taff south of Merthyr Tydfil. This in turn implies that 'Jenkins furnace' is that described as Pontyrhun by a succession of local writers from W. Llewellin (1863a, 88–91) onwards. In 1863 remains of a charcoal furnace could still be seen on the west bank of the Taff opposite Plymouth works but no firm history could be ascribed to the site, apart from the discovery of two sixteenth-century firebacks in the locality. The furnace probably operated in conjunction with a forge about three miles downstream at Pontygwaith (ST 079979). Schubert (1957, 384) was the first to note that the furnace mentioned in a survey of the lordship of Senghennydd of 1625 as being operated by Thomas Erbury was probably that at Pontyrhun, but the site appears to have no later history. It may have been included in the list of 1750–88 simply because remains could be seen close to a furnace then at work or it may have been revived in the eighteenth century. The significance of the name Jenkins is not clear; it is probably coincidental that Llewellin cites the testimony of Mr Jenkins of Cefn Glas farm, Pontygwaith, that a sixteenth-century fireback had been found in the neighbourhood.

On the basis of Llewellin's account of 1863, an antiquity labelled 'Iron Workings (site of)' was published on the OS 1:10,560 map from 1915 onwards at the reference given above. When an officer of the Archaeology Branch of the OS made a field inspection in December 1953, however, he reported that 'No visible remains of the iron works can now be traced' (ex inf. Glamorgan-Gwent Archaeological Trust). No later investigator has located any remains (cf. C. Thomas 1981, 275), which have presumably been washed away by the Taff.

It should be added that the crudely executed watercolour sketch traditionally entitled 'Pontygwaith Ironworks' in the local collection at Merthyr Tydfil Library (Rees 1969, Pl. 35; Hayman 1989, 9) certainly does not depict the site at Pontygwaith in the Taff valley (which was a forge, not a furnace) but could be intended to represent some other local works. The picture shows a square, stone-built blast furnace, with blowing-house, casting-house and charging-house, but no evidence for the method of blowing. Like most other aspects of the early history of ironmaking in the upper Taff valley, the drawing merits further investigation.

Tongwynlais, Glam. ST 129824 [171]

The furnace on the left bank of the Taff at Tongwynlais, close to the boundary between Eglwysilan and Whitchurch parishes, appears to be that operated from 1564 until at least 1568 (but probably not for much longer) by Sir Henry Sidney, in conjunction with a forge on the Rhymney at Rhydygwern, in the parish of Machen, from which bar iron was taken to Cardiff to be shipped to Sidney's steelworks at Robertsbridge in Sussex (Riden 1992b, which reviews earlier accounts of this episode and of the later history of the furnace). Tongwynlais and Rhydygwern are next heard of in a manorial survey of 1625, when they were both in the hands of Thomas Hackett. Nothing further is then known of either works until 1662, when Tongwynlais furnace and Machen forge were leased by John Greenhough of the Van, Caerphilly, to Richard Jones of Hanham (Gloucs.), Richard Hart of Bristol, Gregory Iremonger of Edington (Wilts.) and Alice Stevens of Rhydygwern for sixteen years ending in 1678.

Previous accounts of Tongwynlais have suggested that the site was abandoned about 1680, when a new furnace was built at Caerphilly (q.v.), which was more conveniently situated in the Rhymney valley to supply the forge at Machen. This, however, is to overlook articles of agreement drawn up in 1687 between Sir Charles Kemeys of Cefn Mably and Thomas Morgan of Tredegar for a new lease of the forge at Machen and the furnace at Tongwynlais for a term of 21 years from February 1688. The two works were already in Morgan's hands and the previous tenant is named as Sir Richard Hart, one of the partners in the 1662 agreement. It thus seems likely that Morgan kept on the older furnace for a few years after the new works at Caerphilly (or at least contemplated doing so, since no actual lease appears to survive).

The Tongwynlais site fails to appear in any of the eighteenth-century lists and had evidently been out of use for some time by 1706, when it was described as a 'ruined old Furnace'. In later deeds the property was simply known as Old Furnace Farm, whose position can be located from the

Eglwysilan tithe map. A building still stands on the site, although it is no longer known by this name, and it is possible tentatively to identify a leat that may have carried water from the Taff to a furnace at this spot, which remains unbuilt upon (see Riden 1992b for a fuller discussion).

Ynyscedwyn, Brecs. SN 7809 [160]

A furnace at Ynyscedwyn, in the parish of Ystradgynlais, near the head of the Swansea valley, was apparently in use by 1711, the date on a plate found on the site in 1870 (Schubert 1957, 392); a modern local history (J.H. Davies 1967, 52) refers to a lease of 1696 to Richard Crowley of Old Swinford (Worcs.), which may mark the actual establishment of the furnace, although the document in question has not been located. In 1717 the furnace was listed as 'Uneshadden' and had an output of 200 tons. According to both Minchinton (1961, 11) and Flinn (1962, 13–15) Ynyscedwyn was established by John Hanbury and Ambrose Crowley jointly, but initially made little progress; neither was able to fix a precise date for its building. The Crowley interest continued until at least 1725, when Ambrose Crowley's executor received money on account of Welsh ironworks (he died in 1721).

According to Rees (1968, 307–8) Ynyscedwyn was subsequently leased to John Llewellyn of Ynysygerwyn, Robert Popkin and Richard Seys, but in 1729 Seys and Popkins leased the works to Christopher Portrey of Ynyscedwyn, Henry Williams of Brecon and Gabriel Powell of Swansea. In 1782 the works were in the hands of John Miers, who died in 1786, after which they passed to Richard Parsons, later of Neath Abbey Ironworks, a sequence that fits the evidence of the 1794 list, when Ynyscedwyn was said to be owned by Thomas Aubrey and occupied by 'Mr Parsons'; there was a single charcoal furnace. Richard Parsons also appears briefly (in 1786–7) as a supplier of charcoal-smelted pig from Ynyscedwyn to the Knights' Stour Partnership forges (Ince 1991, 118).

The 1796 list records a yearly output of 1,352 tons at Ynyscedwyn, presumably extrapolated from a weekly figure of 26 tons, which suggests that by this date the furnace (which is marked on Yates's map of 1799) was using coke. In 1806 and 1810 the occupiers were still Parsons & Co.; the furnace was said to be in blast in both years but no output was returned in 1806. The works was much enlarged from the mid-1820s onwards; according to *Mineral Statistics* the furnaces last worked in 1876 (cf. Minchinton 1961, 23–5). A tinplate works was later built on the site, which has since been largely cleared (Hughes and Reynolds 1988, 17–18).

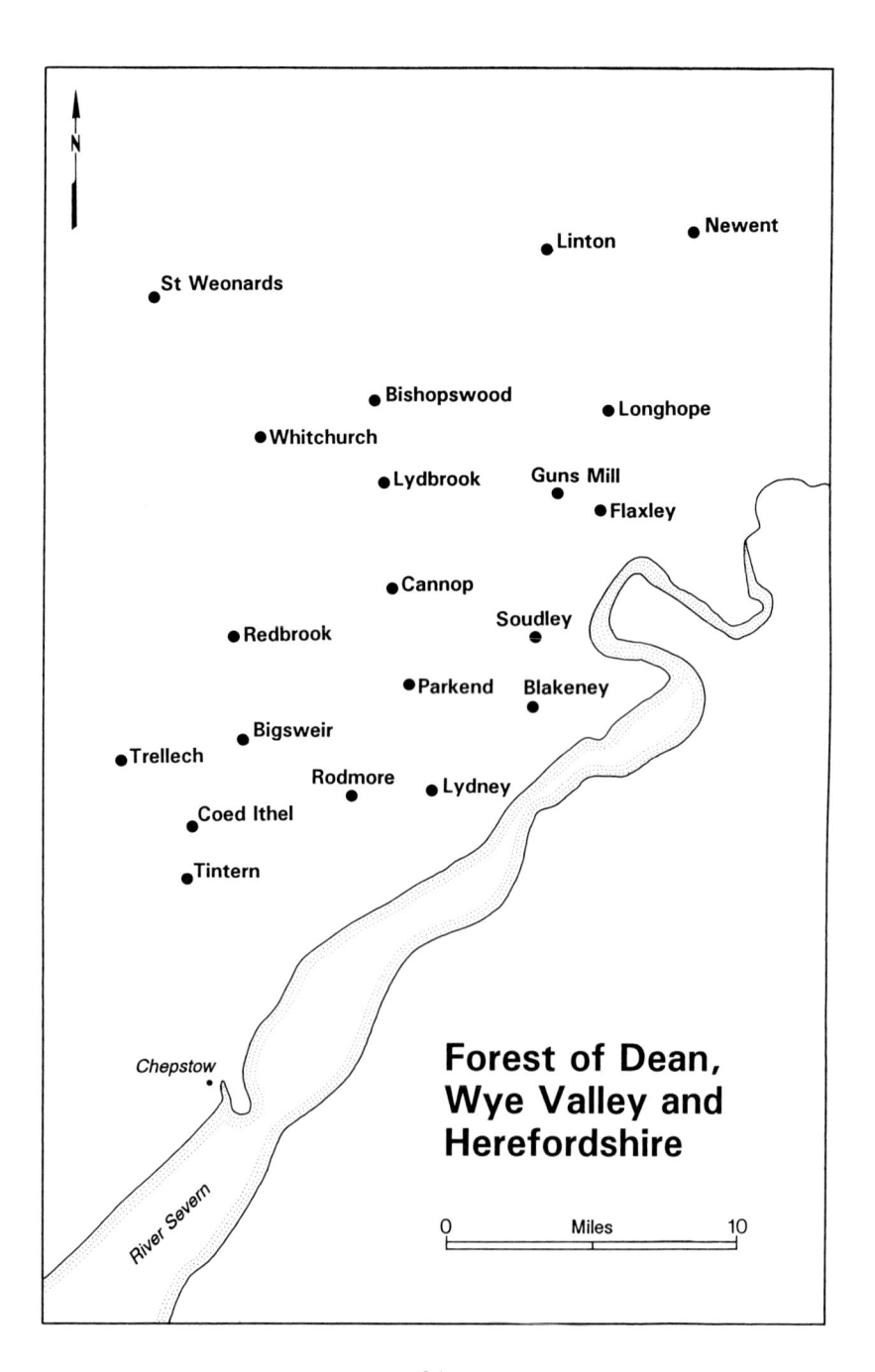

Forest of Dean, Wye Valley and Herefordshire

2

The Forest of Dean, Wye Valley and Herefordshire

Besides the main group of furnaces in the Forest of Dean in west Gloucestershire, this chapter includes three sites in Monmouthshire on the west bank of the Wye and others to the north of the Forest. The early history of blast-furnace ironmaking in this region was dealt with exhaustively by G.F. Hammersley (1971), while for the later period C.E. Hart (1971) collected a good deal of information. The outlying furnaces in Monmouthshire and Herefordshire are also well served by recent work. I am much indebted to David Bick and Ian Standing for sharing with me the results of unpublished fieldwork in this region, where they have both been actively engaged in research for many years.

As elsewhere, the National Grid references in this chapter are followed by the number of the 1:50,000 map on which the reference appears. Anyone exploring this area will in practice find the 1:25,000 Outdoor Leisure Sheet No 14 far more useful, since this includes all the sites described here, except St Weonards.

Bigsweir, Gloucs. SO 5304 [162]

Hammersley (1971, 263) established that references of 1628 and possibly 1625 refer to a furnace close to the Wye near Bigsweir House, the seat of a branch of the Catchmay family, who built the works; they do not, as Jenkins (1925–6, 61, followed by Schubert 1957, 369) believed, refer to Brockweir, lower down the Wye. Similarly, Paar (1973, 36) points out that the reference to Brockweir in 1649 (Suffolk RO, North MSS, cited by Schubert 1957, 369) relates to a shipping-place for iron, not a furnace. The same is apparently true of Brockweir's appearance in the list of Dean furnaces of 1680 (Hart 1953, 103). The conclusions to be drawn from the evidence at present available therefore seem to be (a) there was never a blast furnace at Brockweir, and (b) the furnace at Bigsweir is mentioned only in the 1620s and has no later history.

Bishopswood, Gloucs./Herefordshire   SO 6019; SO 599183 [162]

Hammersley (1971, 261) established that there were two early furnaces in Bishop's Wood, one apparently in Gloucestershire, the other in Herefordshire, which are first referred to in a lease of 1615. Schubert (1957, 368) located references to a furnace here in 1628 and 1639 and noted that Bishopswood appears in the list of Dean furnaces of 1680 (Hart 1953, 103). In 1685 it was in the hands of Paul Foley and was one of the works operated by the family under their partnership of 1692. A broken series of accounts for the Foley works shows that it remained in use until 1751 (Hart 1971, 93–4; Johnson 1953, 135–41). In 1717 output was said to be 600 tons p.a. The Stour Partnership of the Knights was buying pig from this furnace in 1747–9, when the occupier's name was given as either Pendrill or Foley (Ince 1991, 117).

Hart suggests that the Foleys' furnace was then abandoned and that the site is indicated by the names 'Furnace Farm' and 'Furnace Grove' on the Walford tithe map (1840); the first edition 6in. map (1878) marks 'Furnace Wood' and 'The Dam' at SO 6019. The later eighteenth-century references to a charcoal furnace at Bishopswood, he believes, belong to a new works established before 1777 on the Lodge Grove brook about 150 yards from the Wye, where excavations in 1964 located blast furnace and bloomery slag (SO 599183). A forge stood about 300 yards further up the valley. This furnace appears on several maps and is mentioned by tourists in 1770–1, 1777, 1788, 1797 and 1805, on each occasion described as a charcoal furnace. In 1794 it was said to be owned by Lord Foley and occupied by William Partridge; the plant then consisted of a charcoal furnace and finery. In 1796 Bishopswood was said to be producing 500 tons a year, probably meaning 10 tons a week. In 1806 the furnace was listed as being in the hands of William Partridge and producing 650 tons a year. The Knights' accounts show further purchases of pig from Bishopswood in 1772–3, when the tenant was John Mynd, and again in 1788–1801 and 1806–10, this time from William Partridge (Ince 1991, 117).

The furnace is marked on a plan of 1811 but appears to have been abandoned by 1814; the forge remained in use for some years longer. William Partridge built Bishopswood House near the site of the furnace *c.* 1821–2, which in 1826 was described as the seat of John Partridge. It was destroyed by fire in 1873 and the site has now been largely cleared (Hart 1971, 94–6). Schubert's map of Forest of Dean furnaces (1957, 184) marks 'Walford' to the north of Bishopswood, possibly implying that he was aware of the existence of two separate sites, but there is no gazetteer entry under that name.

Blakeney, Gloucs. SO 662068 [162]

The furnace here is apparently first mentioned in 1680 (Hart 1953, 103), when it was in the hands of the Foley family, by whom it was operated until 1715, after which it has no recorded history (Hart 1971, 73–4; Johnson 1951–2, 324, 338). An account of work done at Blakeney in 1692–3 appears to represent either the building of a new furnace or the extensive rebuilding of an existing structure (Bick 1990). In 1717 output was 600 tons and there was an associated forge. The forge does not appear in 1736 or 1750, nor is Blakeney listed in 1794 or in the list of furnaces closed between 1750 and 1788.

The combined efforts of Ian Standing (1986) and David Bick (1990) have located the exact site of the furnace, close to where Taylor's map of Gloucestershire (1777) marks 'Old Furnace', which today is indicated by the name 'Old Furnace Bottom'. In 1984 quantities of slag and charcoal were found between Blackpool Brook and the former Forest of Dean Central Railway at this point. The furnace is marked on a map at the Public Record Office (F 17/7) of *c.* 1700, which also shows a leat running down the valley from a storage pond at SO 653075. Much of the leat has disappeared beneath the later railway but part of the dam wall of the pond can be seen near the main road through the Blackpool Brook valley.

Brockweir: see Bigsweir

Cannop, Gloucs. SO 609116 [162]

A furnace was built on the Cannop brook between Cinderford and Coleford by William Herbert, Earl of Pembroke, in 1612 and was destroyed in 1644. It was among the sites sold for demolition to Paul Foley in 1674 (Schubert 1957, 370; Hammersley 1971, 261). The site is located by Standing and Coates (1979, 17); no surface remains are visible.

Coed Ithel in 1986 after encroaching vegetation had been cut back to reveal the well preserved lining on the two unpierced sides of the furnace. *(S.D. Coates)*

Coed Ithel, Mon. SO 527026 [162]

Excavations in 1964 revealed substantial remains of a seventeenth-century ironworks near the road between Chepstow and Monmouth at Coed Ithel, with a furnace about 24ft (7.3m.) square at the base still standing to a height of some 20ft (6.0m.). It was identified with the reference in a parliamentary survey of Porthcasseg manor in 1651 (Nat. Lib. Wales, Badminton 1631) to a furnace and forge abutting northwardly on the Wye, but this refers to a furnace at Tintern (q.v.). The references to a furnace at Tintern in the 1670s in the Foley accounts certainly do not, as Tylecote suggested, refer to Coed Ithel. There is in fact no clear evidence that the site was used after 1660; the only definite reference appears to be that of 1649 in the North MSS (Suffolk RO), noted by Paar (Tylecote 1966; Schubert 1957, 389–90; Paar 1973).

The site, which is a scheduled monument, lies on private ground on the west side of the main A466 Wye valley road opposite the former home of the Catchmay family, who were probably the builders of the furnace. Requests to view the remains should be made at this house. Parking on the A466 near the furnace is quite impossible; visitors should walk from the nearest safe parking places either in Llandogo village to the north or from a lay-by about half a mile to the south; alternatively, it is possible to descend the extremely steep hillside behind the furnace from the Forestry Commission car-park on the road above. On the site today a substantial section of the furnace itself is still visible, together with some of the walling from the blowing-house and wheel-pit. The well built track which approaches the furnace from the north can also be traced.

Elmbridge: see Newent

Flaxley, Gloucs. SO 693153 [162]

Although there were apparently two early forges at Flaxley, allegedly built in 1627 and 1631 (Hammersley 1971, 263, 264), a furnace seems to be first mentioned only in 1680 (Hart 1953, 103) and appears in the Foley accounts in 1695 and 1710, when on both occasions it was in the occupation of Richard Knight (Johnson 1953, 137; Cave 1974, 19, 22). By 1712 the works (evidently including a forge as well as furnace) were in the hands of Mrs Catherine Boevey; in 1717 the furnace was said to be producing 700 tons a year but no forge is listed. In 1767 the owner was Thomas Crawley-Boevey and in 1794 the works appears with ' — Crawley' as owner and occupier, the plant then consisting of a charcoal furnace, two fineries and a chafery. Flaxley supplied pig to the Knight's Stour Partnership forges over a lengthy period between 1732 and 1786, when the occupier was named first merely as 'Boevey' and later as Thomas Crawley-Boevey (Ince 1991, 117).

In 1796 output was recorded as 360 tons a year, probably a return of actual production; in 1806 the furnace was still in blast, making 379 tons that year, with the occupier listed as T.B. Crawley, although according to Hart smelting at Flaxley ceased in 1802, with the forge remaining in use somewhat longer (Hart 1971, 80–1). Writing in 1802, the local antiquary Thomas Rudge reported that the iron made at Flaxley was 'esteemed peculiarly good', chiefly because both smelting and forging still relied entirely on charcoal as fuel; the furnace was supplied with Lancashire haematite shipped to Newnham, rather than Forest ore (Ibid.).

Both Hart (1971, 81–2) and Ellis (1985) describe the rather slight remains visible at the site of the furnace, which lies in woodland SE of Flaxley Abbey, the home of the owners of the works. Excavations by the Gloucestershire Society for Industrial Archaeology in the 1980s located structures that may have included a wheel-pit and possibly masonry supports for a wooden launder carrying a head-race to the wheel. The site remains part of the Flaxley Abbey estate and there is no public access.

Guns Mill, Gloucs.   SO 675159 [162]

The furnace here was probably built in 1625 by Sir John Winter of Lydney (Hammersley 1971, 262) and is mentioned again 1635 and 1640 (Schubert 1957, 376). The name appears to derive from a William Gunne, living in the early seventeenth century, when the site may have been a corn-mill (Cave 1981); there is no reason to believe that the furnace ever made ordnance. It was destroyed in the Civil War but was operating again by 1682 (Schubert 1957, 376; Hart 1953, 103; Cave 1974, 17–18, 22–24, 29–31). Cast-iron beams over the tapping- and blowing-arches are dated 1683, which was presumably a date of rebuilding, and in 1702 the site was mortgaged to Thomas Foley (Hart 1971, 70). It appears in the Foley accounts (with gaps) between 1705 and 1732 but probably went out of use after the latter date. By 1743 it was a paper-mill (Harris 1974).

Today, Guns Mill is by far the best preserved of the Dean charcoal-fired furnaces and one of the most impressive sites of its kind in England; were the remains to be properly conserved and interpreted the site, a scheduled

Above: **Newent.** An outbuilding at Furnace Farm, apparently converted from the blowing-house of the furnace. *Below:* **Guns Mill** from the south, looking into the casting-arch. Part of the clerk's house can be seen to the left.

monument, would be even more instructive. The furnace itself, which is externally almost complete, is clearly visible from the road running past the site, which is on private property. The furnace was built in front of a natural bank, to which it was connected by a masonry ramp (still intact) from which it was charged. The timber-framed structure which now stands on top of the furnace dates from the period when the site was a paper-mill. The casting- and blowing-arches can clearly be seen on the south and west sides of the furnace respectively, although openings were cut on both sides after the furnace was converted to a paper-mill. There are no significant surface remains of the casting-house but much of the blowing-house is still standing, together with the wheel-pit, in which can be seen the remains of a wheel installed during the paper-mill period.

On the bank behind the furnace, at charging-platform level, the outbuildings of the ironmaster's house to the west of the furnace were presumably once used for charcoal and ore storage. From this point it is also possible to see that the present brick-fronted mansion has been reconstructed from an older stone-built house probably contemporary with the working life of the furnace.

Linton, Herefordshire   SO 661240 etc [162]

A furnace here is mentioned in 1618 in the will of Sir James Scudamore (Taylor 1985–7, 451) and again in the parish register in 1630 (Blick 1987, 68). A deed of 1677 refers to 'the way between the Croose and the Furnace' (Taylor 1985–7, 466, n. 2) and Linton was among the local furnaces listed in 1680. In 1682 Paul Foley paid £700 for the works, which yielded 543 tons of iron in 1686–7 but had gone out of blast a few years later. Although included in the Foley partnership of 1692, Linton had closed down before then and disappears from the partnership accounts in 1698. The furnace was not listed in 1717 or later (Schubert 1957, 380; Hart 1953, 103; Johnson 1953, 134).

Jenkins (1925–6, 62) located the site as lying on the Rudhall Brook near 'Furnace Field' at Hartleton, which now probably lies beneath the M50 motorway. More recently David Bick (1987, 68–71 and pers. comm.) has demonstrated that there are three possible sites in the Rudhall valley, which suggests that there may have more than one furnace known as Linton operating at different dates. Bick draws attention to a site at the bottom of Cut-Throat Lane (SO 666237, also identified by Taylor 1985–7, 466, n. 2), another near Hartleton Farm (SO 649259) and a third at Burton Court (SO

661240), of which the latter seems the best candidate. There appear to be no structural remains of a furnace at any of these, although a barn at Burton Court may have been a charcoal store.

Longhope, Gloucs. SO 686200 [162]

There was probably a furnace here in 1656; it is certainly mentioned in 1680 (Hart 1953, 79–80, 103). It was working in 1682, when Thomas Baskerville saw it on his way from Ross to Gloucester (Hart 1971, 74); in the same year Nourse Yate of Painswick made a lease of the site for 15 years to Thomas Foley. It appears to have no other history and was not listed in 1717 or later. An estate map of the latter year in fact marks a parcel of ground 'called Furnace Close whereon the old Furnes stood' (Cave 1974, 13; D. Bick, pers. comm.; the map itself is in Gloucs. RO, D1297). Bick has located this field at the grid reference given above, about 200 yards NE of Longhope church, where a large quantity of slag is still in evidence. Jenkins (1925–6, 62), on the other hand, identified a site which he called 'Furnace Mill' about a mile lower downstream, which he said had also been used for paper-making and tanning. No authority is cited.

Lydbrook, Gloucs. SO 609152 [162]

Hammersley (1971, 261) dismisses Schubert's suggestion (1957, 380) that the furnace and forge at Lydbrook were operated by the Earl of Essex in 1591–4; a furnace and forge here were among the crown works in Dean in 1613, when there were also three forges in the hands of others. The crown works were destroyed in the Civil War but were later rebuilt and in 1663–4 a furnace called Howbrooke and forges at Lydbrook were being operated by Robert Clayton. The furnace, together with all the king's ironworks in the Forest, were finally abandoned in 1674. The forges at Lydbrook continued in use until the nineteenth century (Schubert 1957, 380; Hart 1971, 74–80). The site of the furnace was located by Standing and Coates (1979, 17) at the confluence of Great How Brook and Lyd Brook; there are no surface remains.

Lydney, Gloucs. SO 628027 [162]

A furnace and forge at Maple Hill (SO 6204), on Cannop Brook north of Lydney, were in existence by 1613 (Hammersley 1971, 262); another forge at Lydney Pill is mentioned in 1628 and a third, again on Cannop Brook near Maple Hill, in 1632 (Ibid., 263, 264).

The early furnace appears to be quite separate from that listed in 1680 (Hart 1953, 103), which was owned and operated, in conjunction with a forge, by the Winter family of Lydney, by whom it was sold in 1723 to Benjamin Bathurst (Schubert 1957, 380–1), whose descendant was named as owner in 1794. Shortly before the sale, the Winters' mortgagees had leased the works to John Ruston of Worcester for 21 years, who surrendered the lease in 1731. Bathurst may have worked the furnace himself for a few years before making a new lease, sometime before 1740, to a Mr Raikes and Rowland Pytt, of whom the latter, in 1731, when he took a lease of the Ynysygerwyn tinplate works near Neath (Glam.), was said to be of Lydney, ironmaster. In 1747 Bathurst made a new lease to Pytt (or his son of the same name) for 21 years.

The younger Pytt died in 1766, demising his lease and other assets to his children through trustees, Francis Homfray of Wollaston, Old Swinford (Worcs.), ironmaster, and John Platt of Monmouth, ironmaster, who took a new lease in 1768. The trustees tried to use coke in the furnace in 1773 but it was 'found not to answer'; two years later they surrendered the lease. The forge was held for a few months in 1775 by Reynolds, Getley & Co., the Bristol ironmasters, but in August that year Bathurst leased all his ironworks for 21 years to David Tanner of Tintern, ironmaster, whose father, an ironmaster of Tintern and Monmouth, had recently died. Tanner remained tenant until he assigned the lease in 1789 to Thomas Daniel, John Fisher Weare, John Scandrett Harford and Thomas Daniel the younger, all of Bristol. The Daniels and their associates held the works only until 1790, when there was a further assignment to John Pidcock, Thomas Pidcock, John Pidcock jun. and Robert Pidcock, 'glassmasters' of Stourbridge, who for a time were in partnership at Lydney with Jeremiah Homfray. The 1794 survey, while correctly listing Bathurst as owner, names the occupier as 'Daniel Harford', which may be a conflation of the names of the lessees of 1789, rather than an otherwise unknown member of the Harford family.

The Stour Partnership accounts confirm the chronology deduced from estate records. Rowland Pytt sold pig from Lydney to the Knights in 1760–63, to be succeeded first by David Tanner in 1776–88 and later the Pidcocks (1792–3) (Ince 1991, 117).

By 1810 the furnace at Lydney had ceased to operate, having in its last years used coke as fuel at least in part, and the forge was advertised to let (Hart 1971, 82–93). In 1717 the furnace was said to produce 250 tons a year; there was also a forge. In 1794 the plant consisted of a charcoal furnace, two fineries, two chaferies, a balling furnace and a rolling- and slitting-mill. The furnace was out of blast in 1796 and does not appear at all in the 1806 list.

The furnace in use from the later seventeenth century until the end of the eighteenth stood to the south of the main A48 on the outskirts of Lydney at the map reference given above; water was brought to the site in a leat which crossed the meadow on the opposite side of the road, supported on stone pillars, and was then carried over the A48 in a stone aqueduct. The tail-race was partly undergound and the entrance to the culvert in which it flowed was located during excavations in the 1980s (Rendell 1986). These investigations also led to the discovery of a furnace bear in the grounds of Whitecross Primary School, which now stands on the site of the works.

Newent, Gloucs. SO 720264 [162]

A furnace at Newent, also known as Oxenhall (in which parish, rather than Newent, the site lies) and Elmbridge, is apparently first mentioned in 1645. It was built by Francis Finch, lord of the manor of Oxenhall, who in 1655 mortgaged the site to Thomas Foley. Three years later Foley bought the leasehold and in 1671 the reversion of the freehold; the furnace remained within the family partnership until 1751 (Schubert 1957, 382; Johnson 1953, 135–6, 141; Styles 1971). As Newent, the furnace appears in the 1717 list but with no output; there were also three forges. Under the same name, and with Foley given as the ironmaster, the furnace appears in the list of those closed between 1750 and 1788. It does not appear, under any name, in 1794. The Knights bought pig from an ironmaster named Pendrill at Newent between 1733 and 1749 (Ince 1991, 117).

Schubert located the site (which he called Elmbridge) as lying on the western outskirts of Newent (1957, 374); Bick (1980; 1987, 62) provides a more accurate description of the remains at Furnace Farm, north of the town, which include a large barn (probably the charcoal store) and a smaller building which appears to have been a blowing-house with charging-platform over. There are no surface remains of the furnace which would have stood in front of the latter. Bick (1980) also points out that while the furnace is marked on Taylor's map of Gloucestershire (1777), two years later Rudder says that it has been 'out of blast for some time'. A number of local cast-iron firebacks, the earliest of which is dated 1643, were almost certainly made here (Bick 1985).

Parkend, Gloucs.   SO 613083 [162]

A furnace and forge were built on the Cannop Brook below York Lodge in 1612 by William Herbert, Earl of Pembroke, which, together with several other works, were destroyed in 1644. In 1653 a new furnace was erected a little lower down the brook, together with a new forge at Whitecroft. Both were abandoned in 1674 and sold for demolition to Paul Foley; they have no later history (Schubert 1957, 383; Hammersley 1971, 261; Anstis 1982, 16–21). The site was located by Standing and Coates (1979, 17); there are no surface remains.

Redbrook, Gloucs.   SO 537107; SO 545108 [162]

A furnace here was in existence by 1613 and is mentioned again in 1628, when it was being worked by the Hall family of High Meadow, who were still there in 1693 (Hammersley 1971, 262; Schubert 1957, 384; Hart 1971, 96). In 1702 Benedict Hall let the furnace, together with two forges at Lydbrook, to Richard Avenant and John Wheeler for 15 years. In 1717 the furnace was said to be producing 600 tons a year and to have an associated forge; at about the same date Redbrook was the scene of an unsuccessful attempt to use coke for smelting (Johnson 1953, 136). The lease to the Foley partnership came to an end in 1725 and the Gage family subsequently built, on or near the same site, a new furnace, which in 1742 was leased to Rowland Pytt, together with two forges at Lydbrook.

Pytt supplied pig from Redbrook to the Knights in 1755–6; he died in the latter year and was succeeded by his son of the same name, who is listed as a supplier in 1761–3 (Ince 1991, 118). In the latter year Lord Gage made a lease of Redbrook furnace and Lydbrook forges to Richard Reynolds of Bristol and John Partridge sen. and jun. (both of Ross-on-Wye) for 21 years. The younger Partridge appears as a supplier of pig from Redbrook in the Knight accounts in 1785–90 (Ince 1991, 118). In 1792 a tourist referred to the extensive works of Harford, Partridge & Co. of Bristol, who appear as Lord Gage's tenants in 1794, when the plant included a furnace, finery, chafery, rolling-mill and balling furnace.

The lease to Harford, Partridge in fact ended in 1793; the furnace (which was out of blast when the survey of 1796 was made) and forges were then let to David Tanner, who was made bankrupt in November 1798, and in 1799 Gage's steward James Davies proposed to take over the lease. By 1802 the furnace was back in blast with Davies as tenant, in partnership with two others and the Gloucester Bank. In 1805 Davies proposed new terms to Gage, which were accepted; the furnace continued until 1816 while the foundry and forge on the same site remained in use for some years more (Hart 1971, 96–104). In 1806 the occupier was listed as Davies & Co., with a single furnace in blast which made 804 tons that year, apparently still with charcoal as fuel.

The site of the furnace is part of a larger complex of former industrial buildings at Upper Redbrook known as 'The Foundry': Standing and Coates (1979, 17) located two furnace sites at SO 537107 approximately SO 545108; there are no surface remains.

Rodmore, Gloucs. SO 582027 [162]

A furnace here was built here in June 1629 by John Powell of Preston and his daughter Eleanor James. Sometime before 1632 it had been let to Sir John Winter (Hammersley 1971, 263). In 1648–50 and 1660 the furnace, together with a forge at Alvington, were worked by a Captain Braine, in partnership with John Gonning, the Bristol merchant (Schubert 1957, 385; Jenkins 1925–6, 63). The forge appears in the 1717 list but not the furnace, which has no later history. It is not referred to in 1680 (Hart 1971, 71). Standing and Coates (1979, 17) located the site at the confluence of the Cone and Aylesmore books; there are no surface remains.

St Weonards, Herefordshire SO 491234 [162]

The earliest definite evidence for a furnace here is a lease of 1661 by the Mynors family of Treago (who still own the site) to William Hall. In 1671 a new lease was made by Robert Mynors to William Hall and Paul Foley. The furnace, which operated in association with forges at Llancillo, Pontrilas and Peterchurch, remained within the Foley empire for the rest of its life but stood outside the main partnership of 1692. It was out of blast in 1677 and in 1678 a suggestion was made that it should be 'retired'. In 1706 a new lease was made for 21 years from 1705 to John Wheeler and Richard

Avenant, which brought St Weonards, but not the forges, into the main Foley organisation. Between 1717 and 1725 it appears to have been outside the group once more and in 1720 William Rea of Monmouth attributed the rebuilding of the furnace to himself. In 1726 the furnace was leased to Thomas Foley and production continued until 1731. It was probably then blown out and finally disposed of by 1736–7. St Weonards appears in 1717 with an output of 300 tons and is also in the list of furnaces closed between 1750 and 1788; the forge remained in use until the 1790s.

The site is today occupied by a farm at the small hamlet of Old Furnace, where a plaque on a later building is inscribed: 'This furnace was rebuilt by William Rea gen. in 1720'; there are no actual remains of the furnace (van Laun 1979).

Soudley, Gloucs. SO 653108 [162]

A furnace and forge were erected on Soudley Brook between Blakeney and Cinderford by William Herbert, Earl of Pembroke, in 1612. They were destroyed in 1644 and the site sold for demolition in 1674 (Hammersley 1971, 261; Schubert 1957, 387; Hart 1971, 72). Standing and Coates (1979, 17) located the site and reported that, although the furnace was largely demolished, some masonry remains were visible on the surface. On the opposite side of the road from the furnace the foundations of a charcoal store can still been (pers. comm. Ian Standing).

Tintern, Mon. SO 529002; SO 513002 [162]

It is now clear that there were two blast furnaces on the Angidy Brook at Tintern at different dates in the seventeenth century, one of which remained in use until the 1820s and has been excavated and conserved.

The earlier furnace stood to the north of Tintern Abbey, immediately downstream from the bridge which carries the road through the Wye Valley over the Angidy Brook, and occupied a portion of the former abbey demesnes known as Laytons. This furnace, together with a forge, was built by the Earl of Worcester; it appears to be first documented in 1629–33 and is mentioned again in 1648 (Courtney 1983, 366, and 1991, 65; Hammersley 1971, 263; Courtney and Gray 1991, 150–2). It is also the site described in a survey of the manor of Porthcasseg of 1651 as abutting northwardly on the River Wye, a reference which has in the past been identified incorrectly with both Coed Ithel (q.v.) and the later Tintern furnace; in neither case does the abuttal fit, added to which Coed Ithel does not lie within Porthcasseg manor. The furnace at Laytons apparently went out of use when the later site higher up the Angidy was built, although the forge remained in use until the nineteenth century and part of the premises are still standing as the Abbey Mill tea-rooms. Nothing remains of the furnace itself.

The furnace higher up the Angidy brook, about a mile west of Tintern Abbey, is apparently first mentioned in the Foley accounts in 1669 and has a well documented history thereafter. It lay on the estates of the Earl of Worcester (Dukes of Beaufort from 1682) and remained in the Foley family's hands until 1688, when it was taken over directly by Beaufort, who in 1699 entered into a five-year partnership with John Hanbury to operate both furnace and forge (Courtney 1991, 66, 68). In 1706 George White jun. leased the furnace and one of the forges in the valley but the following year John Hanbury of Pontypool was said to be tenant of the furnace. In 1739 (nothing being known for certain of the intervening period) White's son Richard leased the furnace and forge and was still in occupation three years later; in 1747–8 Richard White was selling pig to the Knights' Stour Partnership (Ince 1991, 118).

White died in 1752 and was succeeded at Tintern by his nephew Edward Jordan (Llewellin 1863b, 302–3), who was tenant at the time of a comprehensive estate survey in 1763, when Rowland Pytt had the adjacent wireworks and upper forge (Ibid., 315–17). David Tanner, who in 1775 leased the complete ironmaking complex in the valley, was evidently in possession of the furnace a few years before this, since he appears in the Knights' accounts as a supplier of pig in 1771–2 (Ince 1991, 118); he continued to sell pig from Tintern from 1779 until his bankruptcy in 1798, whereupon the works were advertised for sale. A year later Robert Thompson took the lease; he was still there in 1813 (cf. Llewellin 1863b, 318). The Knights' accounts list Robert Thompson as a supplier of pig from Tintern to the Stour Partnership forges in 1800–01; the name William Thompson appears in 1799–1800 (Ince, loc.cit.). In 1821 William Matthews took a lease of the works, which in 1828 he assigned to Copley Brown and Jeremiah Sharp Brown. By this date the furnace had ceased to operate, having been used in its last campaign by David Mushet for experiments with 'wootz' ore from India (Paar and Tucker 1975; Pickin 1982; the former reference does much to disentangle contradictory earlier accounts of the lessees of this site, although Llewellin 1863b remains useful; cf. Tucker and Wakelin 1981 for a tour account of 1782).

Above: **Tintern furnace,** *c.* **1800.** An aquatint of 'Ironworks near Tintern' which, despite considerable licence in showing the abbey in the distance, appears to represent the furnace in the Angidy valley. *(Monmouth Borough Council: Chepstow Museum) Below:* **Tintern today,** viewed from the east, looking towards the blowing-arch, with the wheel-pit beyond the railings to the right.

The second Tintern furnace appears in the 1717 list with an output of 500 tons and in 1794 as being owned by Beaufort and occupied by Tanner, the plant then consisting of an 'old' charcoal furnace, four fineries, two chaferies and a wire-mill. In 1796 an output of only 70 tons (presumably actual production) was listed, while in 1806, when Thompson appears as occupier, the furnace was said to have made 987 tons. Tintern appears, together with several other charcoal furnaces, at the end of the 1825 survey, with a note that the compiler had been unable to obtain any return from these sites, all of which he assumed were out of blast.

The Angidy valley furnace was excavated by John Pickin for Gwent County Council in 1979-80 (Pickin 1982a and b) and has since been imaginatively conserved and interpreted by the local authority. It is freely accessible from the road which climbs up the valley from Tintern, with limited on-site parking. Water from the brook to power the furnace was stored in a large dam (still to be seen, intact, about a quarter of a mile higher up the valley at Tintern Cross) and brought to the furnace in a leat, the last portion of which consisted of a wooden launder supported on stone piers, the bases of which can be seen on the site today, leading to the well preserved wheel-pit. The furnace itself was built into the valley side and about half the structure survives almost to full height. Both the casting- and blowing-arches have gone, but in each case parts of the wing-walls of the arches survive, as do the footings of the walls of both the casting-house and blowing-house, of which the latter, in later years, contained cast-iron cylinders, rather than traditional bellows. Behind the furnace can be seen a clerk's office, with the charging platform above. There are also the footings of a charcoal store.

As the interpretative panels on the site explain, almost the whole length of the Angidy Brook between the furnace and the Wye was dammed for a succession of forges and wireworks. There are few structural remains of these other works, although the ponds are generally well preserved.

Trellech, Mon. SO 491049 [162]

A furnace in Woolpitch Wood, west of Trellech village, appears to have been first noticed archaeologically by the Ordnance Survey in 1958 and was described in general terms by D.M. Rees (1969, 58). It has recently been cleared of trees and access greatly improved.

At present virtually nothing is known for certain of the history of the furnace, which was probably built sometime during the first half of the seventeenth century, possibly by the Probert family of Pant-glas, the large farmstead which still stands to the south of Woolpitch Wood. Trellech may be referred to obliquely in a letter of 1649 concerning the nearby Coed Ithel furnace (Paar 1973), in which case it was then out of use. Nor is it clear when the furnace was abandoned, although it does not appear in any of the nationally compiled eighteenth-century lists. This may suggest that it did not work after 1660, or at least after 1700. If so, the remains described here are if anything more important, since they may belong to one of the best preserved pre-Civil War blast furnaces anywhere in England or Wales.

The furnace stands on the north bank of Penarth Brook towards the eastern end of Woolpitch Wood. A new means of access to the site from the east has recently been created by Gwent County Council, which enables visitors to park at the lay-by on the B4293 about a quarter of a mile to the SW of Trellech village, cross the field on the opposite side of the road and enter Woolpitch Wood at its SE corner, from where a footpath has been built to link with the older Forestry Commission track which runs alongside Penarth Brook past the furnace. This path, built since 1958, has destroyed the front wall of a charcoal store on the hillside above the furnace, of which the other three walls still stand to a height of up to 2.0m. Immediately outside the eastern end wall of the store can be seen a leat coming down the hillside from a storage pond (of which slight traces remain) about 300m. to the north. The track has removed all trace of the final stretch of the leat, which can nonetheless be seen to run directly to the western side of the furnace, which stands on the bank of the brook below, where signs of a wheel-pit and tail-race can be made out.

The furnace itself is about 7.7m. (i.e. 25ft) square at the base, surviving on its western and southern sides substantially intact to an average height of about 3.0m. This structure is surmounted by a smaller upper section, about 1.3m. in height. Both parts of the furnace are faced with mortared rubble. The northern and eastern sides, containing the blowing- and tapping-arches, have collapsed, exposing a portion of square-section lining near the top of the furnace, which protrudes about 3.0m. above the debris. There is no surface evidence for the blowing-house to the north of the furnace; to the east a short length of the southern wall of the casting-house can be seen above ground. The furnace was probably charged from a ramp over the blowing-house, all trace of which was presumably destroyed when the Forestry Commission track was built, although D.M. Rees claimed to have identified evidence of a bridge connecting the charging platform with the higher ground on which the charcoal store stands (*Archaeology in Wales*, 4 (1964), 25).

Walford: see Bishopswood

Whitchurch, Herefordshire SO 5618 [162]

A furnace and forge on the road from Goodrich to Monmouth are mentioned in 1575, when they were owned by George Talbot, sixth Earl of Shrewsbury. Both were in existence in 1613. A furnace, said to be 'newly begun', was built, apparently on the same site, in 1632 (Hammersley 1971, 262). It has also been suggested (Taylor 1985-7, 451) that the furnace in the hands of Sir John Kyrle of Much Marcle, which was the proposed source of pig for Sir John Scudamore's new forge at Carey Mill during negotations between the two in 1627-8, may have been at Whitchurch. The works here were rebuilt again on old foundations in 1657 and the furnace was still in operation in 1672; it is also in the list of Dean furnaces of 1680 but was said to be out of use in 1695. It appears to have no later history and is not in the list of 1717 (Schubert 1957, 390-1; Hart 1953, 103). The name 'Old Forge' may indicate its approximate position.

N

Shrewsbury

Wombridge

Leighton
Coalbrookdale
Kemberton

Kenley
Willey

Church Preen

Abdon
Charlcotte
Hampton Loade

Bouldon

Ludlow
Cornbrook

Bringewood

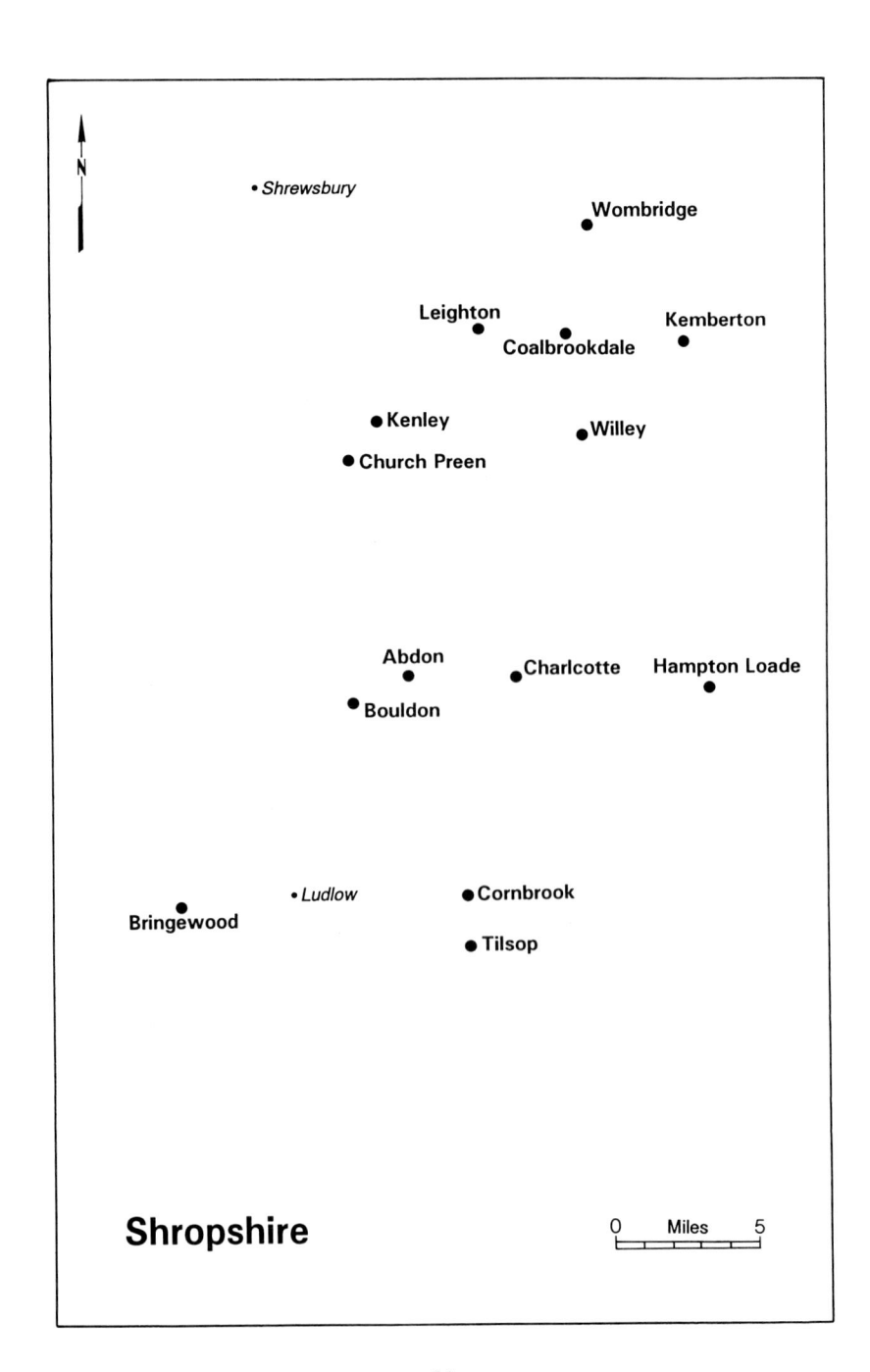

Tilsop

Shropshire

0 Miles 5

3

Shropshire

This chapter includes all the charcoal furnaces in the county which appear to have operated after 1660, apart from Ifton Rhyn, which is grouped with sites in North Wales (Chapter 4), and Halesowen, situated in a detached outlier of Shropshire surrounded by Worcestershire (Chapter 5). Conversely, Bringewood, which was later in Herefordshire, has been included here with the other Clee Hills furnaces. I am grateful to Barrie Trinder of the Ironbridge Institute and Malcolm Wanklyn of Wolverhampton University for help with this chapter.

Abdon, Shropshire SO 567865 [138]

Surviving structural remains at the map reference given above are clearly those of a charcoal blast furnace of whose history nothing at present appears to be known (ex inf. Barrie Trinder). The site, sometimes called Upper Norncott, is not in Schubert (1957), nor apparently in any of the eighteenth-century lists, except possibly that of 1796 (see under Bouldon, below). Malcolm Wanklyn (pers. comm.) is confident that the furnace was built by the Briggs family of Haughton, who were lords of the manor of Abdon in the seventeenth and early eighteenth centuries.

Bouldon, Shropshire SO 5485 [138]

A furnace here is apparently first mentioned in 1644 (Schubert 1957, 368) and appears in the 1717 list with an output of 500 tons. In 1714 it was in the hands of Edward Blount of Teddington (Rowley 1966) but in 1730 was acquired by Richard Knight, who in 1733 passed the site to his sons Edward and Ralph Knight (Page 1982, 12). Bouldon appears in the 1794 list as owned and occupied by Sir Walter Blount, the plant consisting simply of a single charcoal furnace. In 1796 it was offered for sale by Henry Blount, and apparently last worked in 1798, the site later being used as a paper-mill. The list of ironworks of 1796 has an entry for 'Coalford and Bouldon', described as two furnaces, both out of blast. The identity of the first of these names is not clear, unless it refers to the Abdon site, which is little more than a mile from Bouldon, although the name Coalford does not appear on the modern map.

Extensive slag tips can be seen at the Bouldon site today (ex inf. B. Trinder; cf. Rowley 1966 for illustrations).

Bringewood, Shropshire (later Herefordshire)   SO 454750 [138]

According to Schubert (1957, 369) the furnace here was built prior to 1601 and after passing through various hands came to Lord Craven after the Restoration. In 1663 the furnace and forge were leased to Francis Walker, who later assigned the premises to his son Richard. In 1690 Job Walker took a new lease for 31 years; in 1698 Richard Knight obtained an assignment of that lease and two years later acquired the freehold. In 1741 the Knights built a tin-mill at Bringewood but after 1750 the furnace worked irregularly, until its closure in 1778 (Ince 1991, 7–14; Page 1982, 10–14; Lewis 1949, 9).

Bringewood appears in the 1717 list with an output of 450 tons p.a. and in 1794 as a charcoal furnace (a mark in the column listing coke furnaces having been partially erased, presumably as an error), with a finery and three chaferies. The owners were still the Knight family and the occupiers were listed as Downing & Cooley, which corresponds with a surviving lease of 1784 to William Downing of Strangworth Forge, Pembridge, Benjamin Giles of Hope Bagot and John Longmore of Cleobury Mortimer (Ince 1991, 14; Page 1982, 14). In 1796 the output was given as 500 tons a year (probably meaning 10 tons a week), although it is not clear whether there is any other evidence for the continued use of the furnace as late as this. According to H.G. Bull (1869, 54–7), the descendants of Giles held the works until its final closure in 1814 or 1815. Bull also claimed that a furnace was built here as early as 1584.

The site is now indicated by the name 'Bringewood Forge' near Downton Castle, the former home of the Knights. Leats and wheel-pits remain at the site today (ex inf. B. Trinder; cf. Brook 1977, 83). The precise site of the furnace was identified by John van Laun in 1979, who noted that many of the water-power features visible there today belong to landscaping works carried out by Richard Payne Knight in 1817, a couple of years after the works closed (van Laun 1985–7; cf. Bayliss 1985–7).

Charlcotte, Shropshire
SO 638860 [138]

The origins of this furnace are obscure; there certainly appears to be no warrant for J.H. Randall's statement (*VCH Shropshire*, I, 472, copied by Schubert 1957, 370) that in the later seventeenth century it was worked by the Childe family of Kinlet. An abstract of title to Charlcotte manor shows that in 1620 the owner of the estate was Sir Francis Lacon, who conveyed the property to James Grove. It remained in the Grove family's hands until 1678, when it was bought by Dame Mary Yate; in 1712 Apollonia Yate sold to Richard Knight, ironmaster of Bringewood. Two corn-mills are named in the abstract up to 1712, when the furnace is first mentioned. The other mill-site was converted to a paper-mill in the early eighteenth century but when the furnace was established is unclear. The manor was mortgaged to Philip Foley of Prestwood in 1674, who almost immediately made an assignment of the mortgage, but although both Foley and the Lacons were involved in the iron industry, the surviving abstract makes no reference to a furnace before 1712 (Mutton 1966, 1965–8a). In 1733 Richard Knight passed the works to his sons Edward and Ralph and the furnace remained in the family's hands until its closure in 1777 (Ince 1991, 8–12; Page 1982, 11–13; Lewis 1949, 8–9).

Charcotte appears in the 1717 list with an output of 400 tons p.a. and in the list of charcoal furnaces closed between 1750 and 1788, where the occupiers are described as 'Knight & Co.'. Although the furnace was not operated by the Knights after the 1776–7 season, when they sold their Charlcotte property in 1792 the furnace was offered separately, implying that it was either in working order or capable of being made so. It was bought by Thomas Mytton of Shipton Hall but there is no evidence that it ever worked again (Mutton 1965–8a). It was not included in the 1794 list.

The surviving remains of the furnace (a scheduled monument; cf. Brook 1977, 84) can be seen in the field immediately adjacent to the farmhouse named as Charlcotte on the 1:50,000 map; access is from the minor road to the NE of Cleobury Brook and permission should be sought from the house before entering the field. The furnace itself is substantially complete, including well-preserved tapping- and blowing-arches and the bank from which it was charged. The structure was strengthened some years ago by the insertion of concrete beams behind the original cast-iron lintels to the arches but there has been no further conservation or interpretation. There are few if any surface remains of the blowing-house, casting-house or other ancillary features, although there are large quantities of slag near the furnace.

SHROPSHIRE

Above: **Charlcotte.** A general view from the south-east, showing the casting- and blowing-arches. *Below:* **The Old Furnace at Coalbrookdale,** before it was enclosed by the present museum building. *(Ironbridge Gorge Museum Trust)*

Church Preen, Shropshire SO 5498 [137]

In 1727 John Dickins, lord of the manor of Church Preen, announced a scheme to build an iron furnace in the parish, apparently as a speculation to rid himself of debt. It did not go ahead but a prospectus was issued, which gives the venture some interest (*VCH Shropshire*, VIII, 127).

Clee Hill: see Cornbrook

Coalbrookdale, Shropshire SJ 667047 [127]

What later became known as the 'Old' or 'Upper' furnace at Coalbrookdale was apparently built in 1638, the date shown on the tapping-arch lintel, by Sir Basil Brooke of Madeley, unless the beam has been re-used from a steelworks known to have been operated in the Dale at this period by Brooke and his son, also named Basil. There is, in fact, no definite evidence for pre-1660 smelting on the site (Wanklyn 1973; further information from B. Trinder). Prof. K.J. Höltingen (University of Erlangen-Nürnberg, Germany) has reinterpreted and expanded the inscription on the lintel so as to read: 'Brooke Ethelreda [and] Basil 1638 Ethelreda [and] Basil Brooke'; Sir Basil Brooke married Ethelreda, daughter and heiress of Sir Edmund Brudenell, in 1605 (*Ironbridge Quarterly* (1982–3), p. 3). Malcolm Wanklyn (pers. comm.) suggests that the furnace might have been built as early as 1616, when steel-making began at Coalbrookdale. The trustees of the younger Brooke made a lease of the furnace in 1696 to Shadrach Fox, who almost at once assigned the tenancy to William Corfield of Pitchford and Thomas Corfield of Leintwardine.

In 1708 the furnace was leased to Abraham Darby, a Bristol brass- and iron-founder, who reconditioned the plant and in January the following year began a new campaign, using coke as fuel (Schubert 1957, 371; *VCH Shropshire*, I, 460–1; Trinder 1981, 13–16, the latter providing a full analysis of earlier work on the adoption of coke-smelting). Thereafter, Coalbrookdale operated with coke as fuel; a second furnace was added in 1715 (Trinder 1981, 16). The pig was used mainly for casting and only in the 1750s, after further modifications to the process by the younger Abraham Darby, was coke-smelted pig from Coalbrookdale and elsewhere in Shropshire used to produce bar iron. Coalbrookdale appears in 1717 with an output of 400 tons and in later lists as a coke furnace. The furnace was finally blown out in 1818.

The Old Furnace at Coalbrookdale (Brook 1977, 74–5, and Raistrick 1979–80 provide illustrated descriptions) is now conserved and fully interpreted as part of the Ironbridge Gorge Museum, essentially as rebuilt in 1777, although with a lintel over the tapping-arch surviving from the original structure.

Cornbrook, Shropshire SO 6075 [138]

The Knights' accounts show that Thomas Botfield sold charcoal-smelted pig from Cornbrook to the Stour Partnership forges between 1785 and 1788 and coke-smelted pig from the latter year until 1794 and again in 1798–9 (Ince 1991, 117, 119, 127). Cornbrook was the only furnace to sell both types of pig to the Knights: indeed its appearance in the Stour accounts is the first evidence yet noted for the site prior to the 1796 list and the only evidence that it ever smelted with charcoal.

The 1794 list, although making no mention of a furnace at Cornbrook, includes an entry for Clee Hill furnace, also operated by Botfield: the ground landlord was Lord Craven, the furnace was coke-fired, blown by engine and had been built in 1783. Clee Hill is barely a mile from Cornbrook, and it is possible that this is the site, which evidently switched from charcoal to coke in 1788, referred to in the Knights' accounts. On the other hand, Clee Hill (which was not listed in 1796) and Cornbrook are given separately in 1806 and 1810, each with a single furnace. Botfield appears as owner at Cornbrook; Clee Hill was in the hands of George & Co. In 1825 both works were evidently out of use, with 'Wilkinson's Executors' listed as the owner in each case. The suspicion that one site has been mistaken for two is heightened by the remarkably similar annual output figures of 303 and 292 tons returned for Clee Hill and Cornbrook respectively in 1806; in 1796 Cornbrook was claimed to be producing 20 tons a week. Goodman (1978, 377–8) speaks of Botfield building only one (coke-fired) furnace in this area in 1794 or 1795, which operated in conjunction with a forge at Cleobury Mortimer.

The approximate site of the works (or at least one of them) is presumably indicated by the name Old Furnace on Corn Brook to the east of the village of Clee Hill.

Hampton Loade, Shropshire SO 7486 [138]

The list of charcoal furnaces closed between 1750 and 1788 includes Hampton Loade, with the marginal comment 'Forge'. It seems to have no earlier history. The works appears as a forge (only) in the 1794 list, when the owner was Mr Whitmore and the occupier William Jones. The only plant was a single finery. What is presumably the same site was considerably enlarged by John Thompson in 1796–7 and remained in use as a forge until 1866 (Mutton 1965–8b; Trinder 1981, 47, 144).

Kemberton, Shropshire SJ 744044 [127]

Little seems to known of the early history of this furnace, which appears in the 1717 list with an output of 250 tons. It was leased by Robert Slaney in 1714 to William Cotton, Edward Kendall and William Wright, and a few years later may have been the scene of attempts to use coke as fuel. The attempt was evidently unsuccessful, for in the 1720s, when the furnace was described as being in Kendall's hands, pig from Kemberton was being sold for forging and was thus probably charcoal-smelted (Trinder 1981, 15–16; Schubert 1957, 379). Between 1715 and 1717 William Westby Cotton, eldest son of the Yorkshire ironmaster William Cotton, was living at Kemberton, presumably because of the family's interest in the furnace (Awty 1957, 95). The site, which was located by Norman Mutton (1973, 26), appears in the list of closures between 1750 and 1788 but with no additional information.

Kenley, Shropshire SO 574988 [137]

A furnace on the Hughley Brook, SE of New Hall, was erected by Rowland Lacon shortly before 1591, when it was occupied by Richard Holbeck, who was still there in 1606 but moved later the same year to Longnor. John Shaw was the lessee at Kenley in the 1620s and the furnace may have been leased later to William Fownes, who is sometimes described as of Kenley and was a partner at the nearby Harley forge in 1638. Kenley furnace was disused by 1708 and it is not clear whether the site operated after 1660. Fields adjoining the site contain quantities of slag and charcoal (*VCH Shropshire*, VIII, 93, 96).

Leighton, Shropshire   SJ 6105 [127]

The furnace here was built soon after 1630 by Sir Richard Newport of High Ercall on a former corn-mill site and was leased in 1638 to William Boycott and William Fownes. In 1655 Leighton was leased afresh to Boycott, Francis Walker and Joshua Newborough of Stourbridge, the latter an associate of the Foleys. By 1666 it was being operated by a partnership consisting of William Walker of Bringewood, Francis Boycott of Uppington Forge, Lord Newport and Joshua Newborough (the first three being sons of the lessees of 1655) (ex inf. M. Wanklyn; cf. Schubert 1957, 379). From 1681 the furnace was operated by the same partnership as Willey (q.v.). Leighton appears in the 1717 list with an output of 200 tons.

The Willey furnace (and possibly that at Leighton also) was taken over by Richard Ford and Thomas Goldney in 1733, who used coke as the fuel at Willey. About 1756 Richard Ford's sons became bankrupt and the family's interests were acquired by the Darbys of Coalbrookdale. The Leighton works sold shot to the Darbys in 1763 but probably shut shortly afterwards (Trinder 1981, 19, 53); in the list of furnaces closed between 1750 and 1788 it was said to have been converted into a corn-mill, with the owner or occupier given as 'Darby'. A tuyère from the furnace is still *in situ* beneath the later corn-mill, which is now part of the Kynnersley Arms public house (ex inf. B. Trinder).

Tilsop, Shropshire   SO 615725 [138]

The existence of this late seventeenth-century Clee Hills furnace was deduced by K.W.G. Goodman (1980) from scattered indirect evidence. The furnace was apparently built *c.* 1662 or a little earlier and can be shown to have been working in the 1680s. It appears to have been a marginal site, brought into use at times of high demand. The furnace was apparently still working in 1702 but probably closed shortly afterwards. It is not in any of the eighteenth-century lists and no lessees' names appear to be known.

Willey, Shropshire   SO 672981 [138]; SJ 671006 [127]

The furnace here was built after 1609 but before 1618, probably by the Lacon family of Willey, who also had a furnace at Kenley and forges at Harley and Sheinton. It was reconstructed prior to 1631 by John Weld, the purchaser of the Willey estate, but was out of blast during the Civil War (ex inf. M. Wanklyn). In 1674 it was leased to Philip Foley of Prestwood (Staffs.), for 21 years; later the same year he assigned the lease to a partnership which also controlled Leighton furnace (q.v.) (Schubert 1957, 391). This group was evidently made up of Francis Knight, son of the first Richard Knight, together with Richard Baldwin and others, who held Willey until just before 1729, when Richard Knight jun. bought out the interests of his brother, the executors of Richard Baldwin and Sir Richard Smith. Knight held the furnace until 1733 in partnership with his son-in-law Edward Baugh and one of the Payne family (Page 1982, 10–11). Willey was supplying pig to the Knights' Stour forges between 1726 and 1733 (Ince 1991, 118).

In the latter year Richard Ford and Thomas Goldney, acting on their own account and not as partners in the Coalbrookdale Co., took over the furnace, which blew for the first time under the new regime in September 1733 with coke as fuel. This concern continued until 1757, when it was succeeded by a New Willey Co., which included John Wilkinson. The old furnace (SO 672981) had fallen into disrepair but was rebuilt, while a second furnace was built on a nearby site (SJ 671006) in 1758–9 (Trinder 1981, 18–19, 26–6; cf. Brook 1977, 73). In 1717 the furnace was said to be producing 450 tons a year; in the list of charcoal furnaces closed between 1750 and 1788 'Old Wiley' appears with the note 'W Co', presumably meaning Willey Co. (or possibly referring to Wilkinson), although it had apparently been operating with coke since 1733.

Wombridge, Shropshire   SJ 696115 [127]

This may be the 'Rekin' furnace referred to in 1634 (PRO, SP 16/321, ff. 41–70) and was certainly in existence in 1663, when it was leased to Thomas Foley. Wombridge was in production in 1669–70 and included (without any further details) in a list of stock in 1673. It does not appear in a similar list the following year, which may point to its closure. Charlton estate rentals of the 1680s in the Shropshire RO show the works as being out of use, as they do into the early eighteenth century, although the furnace continued to be mentioned. No other history appears to be known (ex inf. M. Wanklyn and B. Trinder; cf.Trinder 1981, ch. 5).

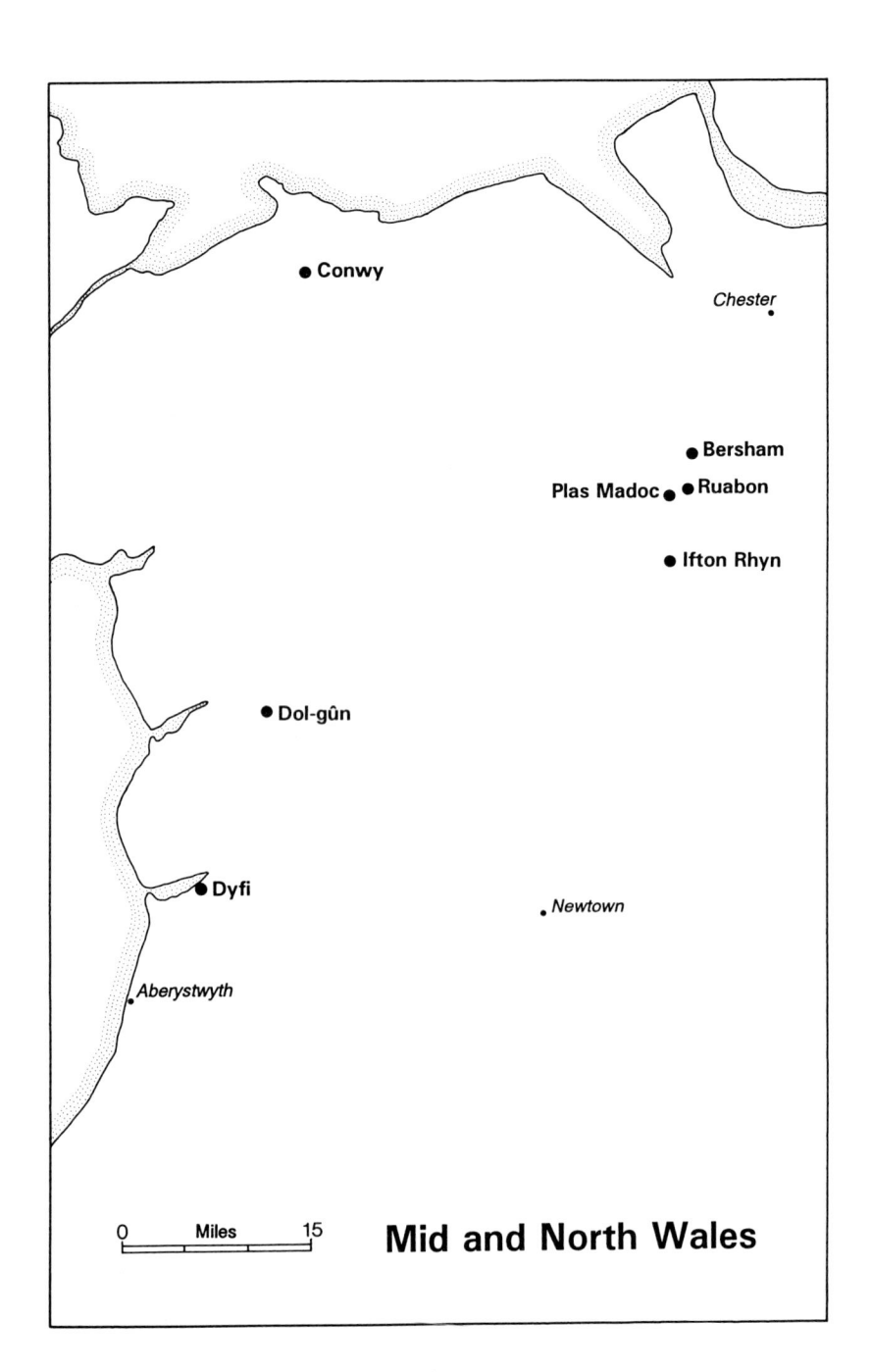

Mid and North Wales

4

Mid and North Wales

This chapter includes the small number of furnaces scattered across Cardigan, Merioneth, Denbigh and Flint, plus Ifton Rhyn in north-west Shropshire, which belongs here rather than in that county. The history of the iron industry in the north-eastern part of this region has been well worked out by Ifor Edwards, whose work forms the basis of most of the entries here. I am also indebted to Stephen Grenter for help in revising the account of Bersham.

Bersham, Denbighshire   SJ 311491; SJ 308493 [117]

Bersham furnace was probably built *c*. 1670 by Sir Richard Lloyd of Esclus, near Wrexham, who died in 1676 (Edwards 1961, 66–80; Davies 1945–7, 84–5). From then until 1690 nothing is known of its history, but from the latter date until 1710 it was managed by Hugh Moore in conjunction with Pont-y-blew forge. Between about 1710 and 1728 it was in the hands of Charles Lloyd of Dolobran, for a time in partnership with Richard Moore, and worked in association with Lloyd's forge at Dolobran. After Lloyd's bankruptcy, Bersham passed into the hands of John Hawkins, Richard Ford and Thomas Goldney; Hawkins died in 1739 and his place was taken by his widow Ann. In 1749 the furnace was sold to Benjamin Harvey and Joshua Gee, whose daughter Sarah married either Harvey or his father of the same name.

In 1753 Bersham was leased by Isaac Wilkinson, previously of Wilson House, Lancs. (q.v.). Charles Lloyd had experimented with coke smelting at Bersham as early as 1721 but, when Wilkinson took over, the works was apparently operating as a charcoal furnace (or with charcoal and coke alternately). According to Palmer (1899), the output at Bersham under Wilkinson was largely castings, including ordnance, which suggests that coke was now being used. The original Bersham partnership apparently consisted of Isaac Wilkinson and his son John, plus some unnamed Liverpool merchants; in 1761–2 the company was reconstructed. According to Palmer, this entailed Isaac Wilkinson's retirement and the continuation of the business by John Wilkinson alone; according to Chaloner (1960, 37-8) it merely involved the exclusion of partners other than the two Wilkinsons. By the 1770s the company consisted only of John and his brother William Wilkinson, who were operating Bersham as a coke-smelting works, with a Newcomen steam engine used to return water. During this period, Bersham became well known as a supplier of engine castings to Boulton & Watt (Edwards 1972).

At some date, apparently 1793, the Wilkinsons' partnership at Bersham came to an end, although the tradition of a violent disagreement between the brothers resulting in the destruction of the works appears to be an invention of A.N. Palmer's (Palmer 1899, 27-8; cf. Dickinson 1914, 37-8). Bersham is described as 'down' in the list of charcoal furnaces closed between 1750 and 1788, with Wilkinson named as operator, and is also included in the supplementary list of coke furnaces abandoned in the same period. In 1794 Bersham was described as owned by Mr Myddleton and occupied by John Wilkinson, with no furnace or forge plant, merely a rolling-mill, built in 1788. Bersham was said in a contemporary appendix to the 1717 list to have an output of 300 tons p.a. After John Wilkinson's death in 1808, his trustees let what remained of the works, which were finally sold in 1812, although the lease of the land on which the furnace stood did not expire until 1828 (Palmer 1899, 37).

The site of Bersham Ironworks is currently being developed by Clwyd County Council as an industrial heritage centre and is marked as both 'Iron Works' and 'Museum' on the OS map on either side on the A483 dual carriageway SW of Wrexham. To the east of the main road a late nineteenth-century school occupies most of the site of the East Works, the main focus of John Wilkinson's activities, where excavations in 1976 (Sale 1978) failed to locate any definite remains of the works. The school itself houses displays on Wilkinson and the Denbighshire iron industry. On the other side of the A483 rather more can be seen of the West Works at Mill Farm, where a number of buildings passed into farm use and have therefore survived. Bersham Mill, for example, was originally built as a foundry in the late eighteenth century, while an octagonal structure nearby was built by John Wilkinson as a cannon foundry about 1775. Also visible on the site are the remains of several other foundry buildings and associated air furnaces; a boiler-pit for a Boulton & Watt engine, possibly used for furnace blowing; the remains of a haystack boiler base and coal bunker, probably connected with Isaac Wilkinson's blowing engine; and a circular casting-pit and crane bed near the furnace casting-arch. In addition, 30m. of wooden waggonway, probably built about 1760 to bring coal and ironstone to the works, was

Bersham. A superbly excavated section of wooden railway, recently discovered at the heritage centre which occupies the ironworks site. *(Clwyd Archaeology Service)*

excavated in 1991. On the other hand, a structure scheduled some years ago as a blast furnace has now been reinterpreted as a limekiln.

The whole of the ironworks is the subject of a continuing programme of excavation by Stephen Grenter for Clwyd County Council, which has also developed an eight-mile trail along the adjoining Clywedog Valley. This contains several other sites of industrial interest, for which there is interpretative material on display at Bersham.

Conwy, Denbighshire SH 7972 [115]

A furnace at Talycafn, in the grounds of Bodnant, three miles south of Conwy (Caernarfon), but just inside the Denbighshire parish of Eglwysbach, was built about 1748 by William and Edward Bridge; Ralph Vernon and the Kendall family apparently also had an interest in the furnace, which drew in part on supplies of Furness haematite. Charcoal for the furnace was carried on the River Dee, but this traffic largely ceased after 1767. The Knights' accounts show that 'Bridge & Co.' sold pig from Conwy to the Stour Partnership forges in 1763–4, as did one of the Kendalls in 1767–71. The Bridge brothers were bankrupt in 1773 and in July 1774 both Conwy and Dyfi (q.v.) furnaces were advertised for sale. No buyer was found for either but Conwy was offered again in 1779. It then seems to have been abandoned (Schubert 1957, 372; Awty 1957, 112, 115, 122; I. Edwards 1961, 63–6). In the list of closures between 1750 and 1788 Conwy appears with 'Bridge' as occupier and the comment 'stands'; it does not appear in the 1794 list.

Dol-gûn, Merionethshire SH 751187 [124]

A furnace at Dol-gûn, about a mile east of Dolgellau, was projected by Abraham Darby in 1713 but construction only began in 1717, the year of his death. John Kelsall left Dolobran forge the same year to become clerk at the new furnace, which was sold to Samuel Milner, another member of the Society of Friends, in 1717. Kelsall left Dol-gûn after the first campaign in 1719 and returned to Dolobran. Between then and 1732 there were eight short campaigns at the furnace, which in 1729 was sold to Henry and John Payton, Quakers from Dudley (Worcs.), who brought Kelsall back as clerk. In 1734 the furnace was blown out and Kelsall departed for Ireland (Schubert 1957, 373–4; Raistrick 1968, 125–8). Thomas (1981–4; no authority cited) adds the additional detail that from 1729 the use of local ore was given up in favour of haematite from Furness and that about 1720 a forge was built between the furnace and Dolgellau. Afon Wnion was navigable up to Dol-gûn and boats were used to ship finished iron out to the Mawddach estuary.

Although Schubert says that the furnace was abandoned 'by 1771', there seems to be no evidence that it operated after 1734. Thomas (loc.cit.) gives a date of 1743 without citing a source; this may be a typographical error for 1734. Dol-gûn is not in any of the eighteenth-century lists and Schubert's date evidently derives from A.S. Davies's reference (1945–7, 86) to an unsuccessful patent application by Edward Thomas of 'Dolgyn Forge' and John Chadwick of Duxbury (Lancs.) for smelting with peat in 1771, which points to there being a forge only by this date. But Davies also cites an advertisement in the *Salopian Journal* (5 May 1802) for the sale of 'Dolgyn Furnace', including 'a blast engine worked by water power', a forge and charcoal, with the capacity to make 6 tons of bar or 8–10 tons of half-blooms weekly, from which it is unclear whether the site was revived.

Substantial remains of the furnace have been excavated by Peter Crew for the Snowdonia National Park (Blick 1984, 47–8; cf. Thomas 1981–4 for further topographical detail).

Dyfi, Cardiganshire   SN 685951 [135]

The furnace on the River Dyfi in the parish of Eglwysfach is sometimes called 'Aberdovey', despite being situated several miles from the coast; the nearest town is Machynlleth. It was built in 1755 by Ralph Vernon and the Kendalls, a few years after their venture at Conwy (q.v.); as at Conwy, the brothers Edward and William Bridge were apparently also involved (Awty 1957, 112; Edwards 1961, 65–6). Pig from Dyfi was being sold to the Knights by the Bridges in 1763–4 and by Kendall & Co. in 1765–7 (Ince 1991, 117). In the 1780s pig was also being sent to a forge at 'Llanfrede' (i.e. Glanfraid, SN 6387) in the Leri valley a few miles south of the furnace (*Archaeology in Wales*, 10 (1970), 30).

Like Conwy, Dyfi furnace was advertised for sale in 1774, following the Bridges' bankruptcy, but unlike the other site, which was abandoned after failing to find a buyer, Dyfi was kept on by the Kendalls. According to the list of closures between 1750 and 1788, Dyfi furnace was 'down', but it reappears in 1794 with Lewis Edwards as owner and the Kendalls as occupiers. There was a single charcoal furnace, apparently water-blown, and no other plant. In 1796 output was returned as 200 tons a year, probably its

Dyfi, from the north-west, showing the restored furnace and blowing-house, with charging-bridge over; the casting-house stood on the opposite side of the furnace. To the right can be seen Tŷ Furnace, the clerk's home, and to the left the charcoal store. (*Crown Copyright: RCAHM Wales*)

actual production; Kendall & Co. are to be found in the Knight accounts in 1796–7 selling pig from Dyfi to the Stour forges (Ince 1991, 117). Walter Davies, the Welsh reporter for the Board of Agriculture, noted that in 1797 a forge at Machynlleth was using pigs smelted at Dyfi, which was itself supplied with haematite from Furness (D.M. Rees 1964–7, 118–21; cf. M.I. Williams 1973–4).

An unpublished history and description of the furnace (Griffiths 1953?) notes an advertisement in the *Salopian Journal* (3 June 1795) in which a 'Mr Bell' is mentioned, and a statement by the county historian, Sir Samuel Meyrick, in 1810 that the furnace was not then worked. A tour account of 1805 describes the furnace as abandoned four or five years earlier, following Bell's death, and adds that the building was now rented as a woollen manufactory. Bell was evidently an agent for the proprietors, rather than principal (ex inf. Peter Wakelin). Although the furnace reappears in the survey of 1806, still in Kendall's hands, with an output of 150 tons a year, this may have been an inspired guess on the compiler's part, rather than the result of actual enquiry. The *Cambrian Guide* (1813) also describes the furnace as idle; D.M. Rees's date of closure of 1814 is probably a little too late.

The hamlet in which the furnace was built is now known as Furnace, where the structure survives largely intact as a guardianship monument and has been extensively excavated and conserved by Cadw (L. Griffiths 1953? and D.M. Rees 1964–7 for early accounts; Dinn 1988 describes recent work there). Dyfi thus joins Duddon (Lancs.) and Bonawe (Argyllshire) as one of the three best preserved and interpreted charcoal blast furnaces in Britain. The furnace itself, which measures 9.1m. square at its base, decreasing to 8.0m. at the eaves, 10.1m. above floor level, is substantially complete, with brick-arched blowing- and casting-arches. Inside, the stack is circular in plan, with a well-preserved lining above the boshes and remains of the hearth below. It is also possible at Dyfi (uniquely among the major conserved sites) to see a water-wheel of roughly the appropriate size (9.0m. diameter; whereas earlier wheels were apparently between 3.6m. and 7.9m. in diameter) alongside the blowing-house and charging-ramp. On the opposite side of the furnace part of the walls of the casting-house survive as footings. On the higher ground above the main working area stands the usual large charcoal store. Tŷ Furnace, on the roadside south of the bridge over the River Einion, was once the home of the clerk in charge of the works.

Ifton Rhyn, Shropshire   SJ 296372? [126]

I. Edwards (1957–60; cf. Schubert 1957, 378) outlines the history of a furnace operated by the Myddleton family of Chirk Castle at Ifton Rhyn, just inside Shropshire, from the 1620s. It does not appear to have any history beyond the death of Sir Thomas Myddleton, whose will of 1666 mentions Ifton and Ruabon (q.v.) furnaces. The site may have been at the grid reference given above (ex inf. I. Edwards).

Plas Madoc, Denbighshire   SJ 287436 [117]

Little appears to be known of this site, although part of the structure survives near the site of Plas Madoc Hall, Ruabon (I. Edwards 1961, 85–90, 95, and pers. comm.; Schubert 1957, 384). It is apparently first mentioned in 1677–8 when it was operated by Edward Lloyd of Plas Madoc, who died in 1691. It remained in the hands of the Lloyd family, who appear to have experimented with coke-smelting there in 1757–61 and possibly also the 1740s. They certainly supplied charcoal-smelted pig to the Knights between 1744 and 1751 (Ince 1991, 118). In 1717 the furnace was said to have an output of 300 tons p.a. A daughter of the Lloyd family married John Rowland, whose family continued to be associated with ironsmelting in Denbighshire until the 1820s. Plas Madoc appears in the list of closures between 1750 and 1788 with 'Rowlands' as occupier and the comment 'down' (as at Ruabon, q.v.); it is not in the 1794 list.

Ruabon, Denbighshire   SJ 290458; SJ 304435 [117]

The history of the furnace at Ruabon can be traced from 1634, when it was operated by the Myddletons of Chirk Castle (I. Edwards 1961, 81–4; Schubert 1957, 386; A.S. Davies 1945–7, 85). Between 1662 and 1675 William Cotton was associated with the Myddletons at Ruabon; its history is then a blank until 1693, when it was leased by Thomas Lowbridge and Richard Knight, who remained there until 1710. Between 1722 and 1731 William Wood and Thomas Harvey operated Ruabon in conjunction with Tern forge, Shropshire; between 1731 and 1735 it was in the hands of Daniel Ivie and another member of the Wood family, William Wood having died in 1730 and Harvey in 1731.

Edwards was unable to trace the history of the site after the mid-1730s, apart from a stray reference to William Higgons being in possession in 1763, and suggested it may have closed about 1735. The 1750–88 list, however, describes Ruabon as 'down', but gives the occupier as the Rowland family, who also had the nearby Plas Madoc furnace (q.v.). In 1717 output was said to be 250 tons p.a. In 1790 (according to the 1794 list) a firm named Jones & Rowland built a coke furnace at Ruabon on land belonging to Richard Myddleton, which marks the Rowlands' switch into coke-smelting, presumably having abandoned Plas Madoc and Ruabon in the previous few years (cf. I. Edwards 1965, 167–8). Edwards (pers. comm.) notes two separate sites for charcoal furnaces at Ruabon (as above); the coke furnace was at SJ 293449.

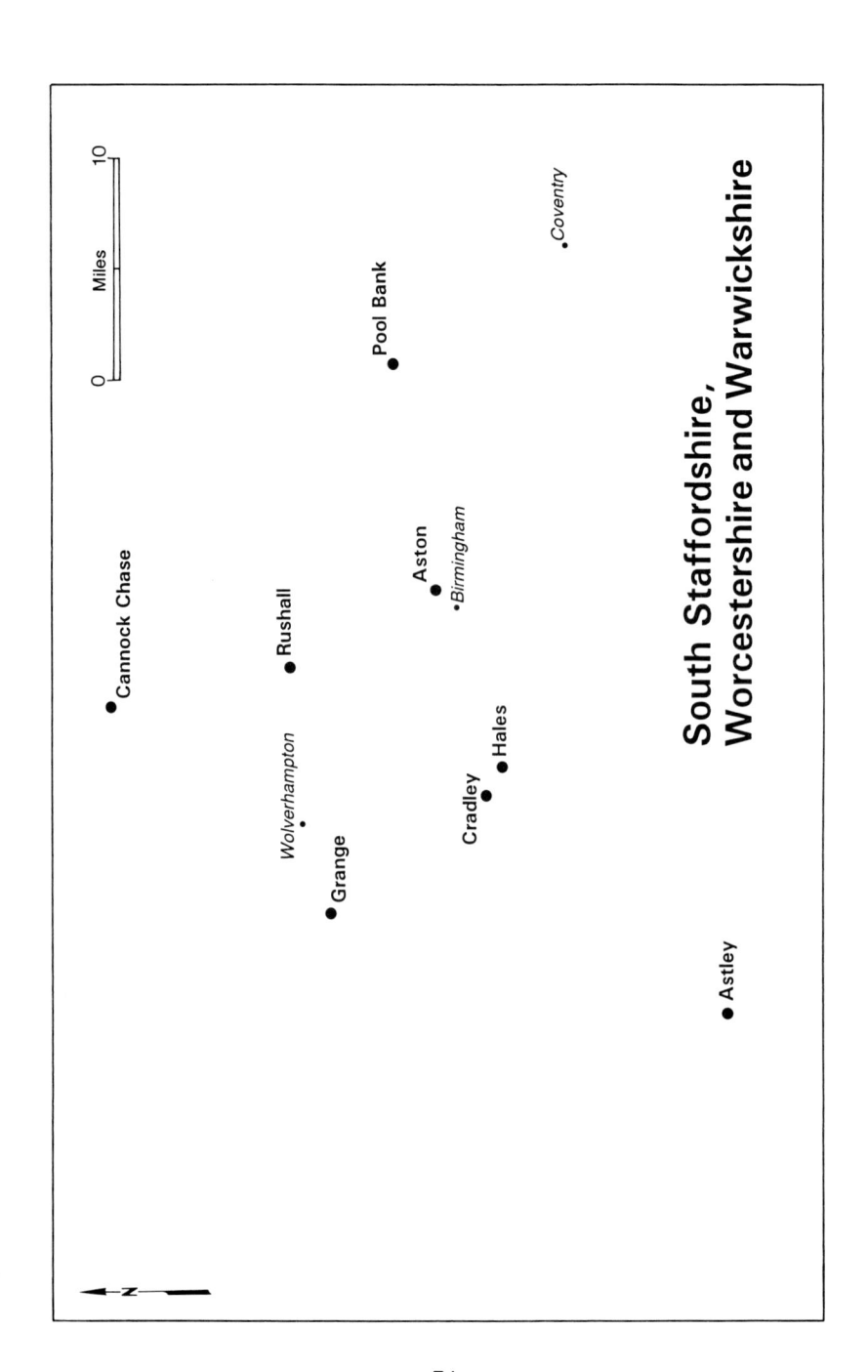

South Staffordshire, Worcestershire and Warwickshire

5

South Staffordshire, Worcestershire and Warwickshire

Astley, Worcs. SO 794665 [150]

A furnace on the Dick Brook NW of Worcester has attracted considerable attention since its rediscovery following flooding in 1924, mainly because of its supposed association with the seventeenth-century projector, Andrew Yarranton (1616–84). A detailed modern survey of both the physical and documentary evidence (Brown 1982, updating Cantrill & Wight 1930; cf. Schubert 1957, 387, and Hallett and Morton 1968) describes and illustrates the surviving remains but was unable to confirm the connection with Yarranton. Nor has any definite chronology for the site been established: the traditional date of construction of 1652 is only tenable if the unproven link with Yarranton is also accepted, although Brown has drawn attention for the first time to a pamphlet, probably written in 1662, which refers to Astley furnace receiving 500 loads of charcoal a year for the last six years, with which to smelt local cinders. No later references to the site have yet been noted; it does not appear in any of the eighteenth-century lists.

Aston, Warwickshire SK 069890 [139]

A furnace on the Hockley Brook, near the junction of Porchester Street and Furnace Lane, was built prior to 1615 by Sir Thomas Holte of Aston Hall on his own estate. In the 1630s it was being worked by Richard Foley (Schubert 1957, 366) but by 1651 was in the hands of John Jennens, who bequeathed it (in a will proved in 1653) to his son Humphrey Jennens (Jennens 1879, 144–5). From the 1630s Aston operated in conjunction with Bromford forge and in the early eighteenth century both works were tenanted by Christopher and Riland Vaughton, and later John Mander and Phelicia Weaman. In 1746 Aston and Bromford, together with a slitting-mill in Nechells Park, were leased to Abraham Spooner and Edward Knight (Ince 1991, 20; *VCH Warwick*, VII (1964), 259; Lewis 1949, 25; Fairclough 1984,

19–20). The furnace remained within the Knights' Stour Partnership until its closure in 1784–5, supplying pig to forges in Worcestershire. In 1717 Aston was said to produce 400 tons a year; it appears in the list of furnaces shut between 1750 and 1788 with the additional information that it had been worked by Spooner and was now a 'B Mill', presumably meaning brass-mill, although according to *VCH* (loc. cit.) the site became a paper-mill in the early nineteenth century. All trace of the works has disappeared.

According to Dent (1880, 339–40), a Newcomen engine was installed in the later eighteenth century to work the bellows; if so, Aston must have been one of the very few charcoal furnace to have been steam-blown (unless the engine was used simply to return water from the tail-race to the pond). The furnace closed too early to appear in the 1794 list which normally states whether plant was water- or steam-powered.

Cannock Chase, Staffs. SK 008135; *c.* SK 008136 [128]

There were three early furnaces on Cannock Chase, all operated in the mid-sixteenth century by the Paget family, until the third Lord Paget fled to the Continent in the 1580s after a series of religious and political indiscretions. Those known as the 'Old' and 'New' furnaces are generally described as being at 'Cannock Wood'; the third was on the Littleton estate on the western edge of the Chase. The use of the name 'Cannock Wood' can cause misunderstanding: this term is used historically of the whole Chase and should not be confused with the modern settlement of the same name. (This and the next paragraph are based on information kindly supplied by Dudley Fowkes, the Staffordshire County Archivist; cf. Schubert 1957, 370; Morton 1964–5 and 1966.)

The site of one of the 'Cannock Wood' furnaces, presumably the New Furnace, is shown on the county maps of both Smith (1747) and Yates (1775), even though it had probably long been disused. Both the Old and New furnaces in fact lay in the parish of Rugeley, north of Hednesford and west of Beaudesert Old Park (SK 008135); 'Furnace Coppie', just outside Beaudesert Park, was the source of much of the cordwood for these sites. None of the furnaces appears in the nationally compiled list of 1717, nor in the local list of 1735, except as a forge, but a furnace, again presumably the New Furnace, was leased in 1732 to Edward Hall (Awty 1957, 99–100; the lease is Staffs. RO, D603/E/1/826–827), when it was described as a 'furnace now demolished', 'late used' by Hall, Warine Falkner, Thomas Cotton and Edward Kendall. Hall was given power to rebuild the site but it is doubtful whether he did so, despite the appearance of the furnace on the two county maps, or even whether the furnace had been much used by the previous lessees.

Morton (1965a, 319) located the pool and a hearth at Teddesley (SK 008136 or thereabouts), which he clearly distinguished from the Cannock Chase furnace described by Brook (1977, 111).

Cradley, Worcs. SO 9385 [139]

A furnace on the Stour at Cradley, near Halesowen, is first mentioned in 1610, when it was owned by Lord Dudley and operated by Humphrey Lowe. By 1636 it was in the hands of Richard Foley; in 1662 it was leased to Thomas Foley of Great Witley (Schubert 1957, 372). In 1724 Edward Kendall of Stourbridge acquired the lease of Cradley ironworks, previously held by the Wheeler family, and from 1728 the furnace was one of several in the Birmingham area supplying pig to the Knights' Worcestershire forges, as it continued to do until at least 1748 (Awty 1957, 98; *VCH Staffs.*, II (1967), 120; Ince 1991, 117).

The Kendalls' connection with the works probably ended *c*. 1746–51 (Awty 1957, 114) and in 1774 the Dudley estate made a new lease of the forge and furnace at Cradley to William and Richard Croft for 21 years (Dudley Archives, D/DE IV 3, formerly Box 15, Bundle 8). The furnace appears in the list of closures between 1750 and 1788 with the name 'Crofts' as operator, and the marginal comment 'forge'. It is omitted from the 1794 list, although two forges, leased by William Gibbons & Co. from the Dudley estate, are included at Cradley and Lye. This corresponds with the evidence of a later lease (Ibid., Box 16, Bundle 13) of 1792 for 21 years to Thomas, William and Benjamin Gibbons, which recites an assignment of the previous lease by Croft to Gibbons (the date of the assignment not being stated), and grants a new term to the Gibbonses in Cradley forge and a building 'heretofore used as a Furnace for Pig Iron called Cradley Furnace'. From an examination of Dudley estate rentals it would seem that the assignment took place between 1784 and 1789 and that these years may have seen the closure of the furnace and the conversion of the site to a forge (ex inf. Dudley Archives). On the other hand, Yates's map of Staffordshire of 1775 marks only a string of forges in the Stour near Cradley and Lye, so that conversion may have taken place immediately after the 1774 lease. In 1717 Cradley was said to have an output of 200 tons a year.

Grange, Staffs. SO 845965 [138]

Like others in the Birmingham area, the furnace at Grange, on the Smestow brook in the parish of Penn west of Wolverhampton, is first mentioned in 1636, when it was in the hands of Richard Foley (Schubert 1957, 376). Nothing else seems to be known of its history until the early eighteenth century, when it appears in the 1717 list with an output of 450 tons p.a. and is the only blast furnace included in the survey in south Staffordshire. Between 1728 and 1753 Grange appears in the Stour Partnership accounts of the Knight family as one of several independent furnaces supplying pig to their forges (*VCH Staffs.*, II (1967), 120). It was included in the list of closures between 1750 and 1788 with the name 'Jordan' as operator and the marginal comment 'down'; Jordan is also the tenant's name in the Stour accounts between 1747 and 1753 (Ince 1991, 117). The site, which is not in the 1794 list, is marked on Yates's map (1775, apparently the latest evidence for its continued use) and was located by Morton (1965a), who found slag and charcoal impregnated soil there. The name Furnace Grange on the modern map indicates its position.

Hales, Shropshire (later Worcs.) SO 966845 [139]

A furnace at Halesowen, known as Hales, apparently replaced bloomsmithies there between 1602 and 1606. It was built by the Lyttelton family of Frankley and was operated initially by Humphrey Lowe (Schubert 1957, 376). In 1631 Hales supplied hammers and anvils to Sir John Scudamore's new forge at Carey Mill (Herefordshire) (Taylor 1985–7, 460–1).

Between 1692 and about 1705 the furnace was in the hands of the Foley family. It was later operated by Sir Thomas Lyttelton of Hagley Hall and partners, who were joined in 1726–7 by Richard Knight and his son Edward, the partnership thereafter consisting of Lyttelton, the Knights, Joseph Cox and the executors of Clement Acton. Hales thus became part of the Knights' Stour Partnership and remained in the family's possession until its closure in 1772 (Lewis 1949, 25; *VCH Staffs.*, II (1967), 120; Ince 1991, 3–4, 20). The furnace appears in the 1717 list with an output of 500 tons p.a. and in the list of those closed between 1750 and 1788 as being operated by Knight & Son, now converted to a rolling- and slitting-mill. In the 1794 list there are two works owned by Lord Westcote (as the head of the Lyttelton family became in 1776) in Halesowen: a slitting-mill operated by Attwood and a chafery and balling furnace occupied by Ward; the former must be the successor of the Knights' furnace.

The site (cf. Brook 1977, 183–6) was at the foot of Furnace Hill, in that part of Halesowen parish which until the early nineteenth century was a detached portion of Shropshire in north Worcestershire; it therefore appears under the former county in contemporary lists.

Pool Bank, Warwickshire　　　　　　　　　　　　　　SP 244913 [140]

A furnace known as Pool Bank in the parish of Over Whitacre, west of Nuneaton, whose site is approximately indicated by the modern settlement of Furnace End, was in the hands of Humphrey Jennens of Erdington in 1690, when it was bequeathed to his son John (Jennens 1879, 117–21, 176–7). It is not mentioned in the will of Humphrey's father John Jennens, proved in 1653 (Ibid., 141–52). Pool Bank appears in the 1717 list with an output of 300 tons but is not in either the 1794 list or the list of furnaces closed between 1750 and 1788. It is marked on Henry Beighton's map of Warwickshire (1725) but has no later history.

Rushall, Staffs.　　　　　　　　　　　　　　　　　SP 018997? [139]

This furnace was omitted from the survey of 1717 but between 1725 and 1728 supplied pig to the Knights' Stour valley forges; it is also included in a local list of Staffordshire furnaces of 1735 (*VCH Staffs.*, II (1967), 120; Ince 1991, 118). The tenant in the 1720s may have been Edward Kendall of Stourbridge (Awty 1957, 98). The furnace, about a mile and a half NE of Walsall, appears in the list of closures between 1750 and 1788, where it is said to have been converted into a corn-mill. In 1742 James Bourne of Rushall furnace was co-lessee of a forge at West Bromwich (*VCH Staffs.*, XVII (1976), 33); the furnace is also mentioned in a deed of 1748 (Staffs. RO, D948/1/1/8). An advertisement in *Aris's Birmingham Gazette* (25 April 1768: Morton 1965a, 320) announces the sale of the remaining term of a lease of Rushall furnace (together with Hints Forge, near Lichfield), formerly held by John Churchill of Hints, bankrupt and deceased. The furnace is not marked on Yates's map (1775), nor is it in the 1794 list; Yates, however, marks a mill-site at approximately the reference given above that might be the former furnace.

Cheshire and North Staffordshire

6

Cheshire and North Staffordshire

Disley, Cheshire   SJ 9784 [109]

A furnace near the boundary of Yeardsley Whaley, on the estate of the Jodrell family, is apparently first mentioned in the 1690s, when small quantities of pig from there were received at forges in south Yorkshire. In 1702 some reached Carburton (Notts.). In 1737 the furnace was in the hands of Samuel Bagshaw; in 1777 the tenant was Joseph Lowe (Awty 1957, 89–90). Nothing else appears to be known of the site, which is not in the lists of 1717 or 1794, nor, except perhaps as 'Lea' (with no further information), in the list of closures between 1750 and 1788. It was not included in Schubert's gazetteer.

Doddington, Cheshire   SJ 7147 [118]

A furnace at Doddington is mentioned in 1667 as being owned by Sir Thomas Delves, but little is known of its history. It appears as 'Doddington Works' in the Foley accounts in 1710–11, when it was acquired by the partnership, and later passed to the Kendall family. Hall & Co. supplied pig from Doddington to the Knights' Stour forges in 1749–51 and 1757–8, and sales were continued by Kendall & Co. until 1762 (Ince 1991, 117). A list of Cheshire ironworks prepared in 1766 by an advocate of the Trent & Mersey Canal project includes Doddington furnace and the site is shown on Burdett's map of the county of 1777 (Awty 1957, 78, 92, 113, 116; Schubert 1957, 373). The furnace appears in the list of closures of 1750–88 as 'Derrington' (which is almost certainly not the small settlement of that name near Stafford), with the occupier as Kendall and the marginal comment 'down'. In 1717 it was said to have an output of 500 tons p.a.; it is not in the 1794 list.

Heighley, Staffs.   SJ 7747 [118]

The Royalist Composition papers show that Walter Chetwynd of Rugeley was in possession of various Cheshire and north Staffordshire ironworks in the mid-seventeenth century, including a furnace at Heighley, which was probably demolished by 1680 (Awty 1957, 71, 91). The furnace was not noted by Schubert (1957), nor does it appear in any of the eighteenth-century lists, and seems to have no other recorded history.

Horton: see Leek

Lawton, Cheshire   SJ 8056 [118]

The furnace here was built in 1658 by John Turner, a Stafford ironmonger who also operated Lizard forge in this period, and in the 1660s was smelting Furness haematite. Probably between 1680 and 1687 it came into the hands of William Cotton of Haigh, near Barnsley (Yorks.), and was managed for him by his cousin Thomas Hall. After 1696 Lawton was managed for the Cheshire group of ironmasters by Daniel Hall, who was replaced in 1706 by Daniel Cotton. The furnace is included in the 1717 list with an output of 600 tons but has no later history until it appears as 'the old Furnace at Lawton' in a lease of 1744, when it was converted to a flint-mill (Awty 1957, 74–5, 77, 84–6, 93, 96, 112; Lead 1977, 3, 7; Schubert 1957, 379). In the list of closures between 1750 and 1788 the furnace appears as Lawton Gate, bracketed with Madeley (Staffs.) and Meir Heath as being operated by 'Mr Cotton' and said to be 'down'.

Leek, Staffs.   SJ 9457 [118]

A furnace at Horton, three miles west of Leek, is apparently first mentioned in 1719. It was in the hands of William Fallowfield in 1729 and 1731, when attempts were made to use peat as fuel, and in 1735, when it appears in Wilkes's list of Staffordshire furnaces. It seems to have no later history and may have shut in the depression in the iron trade of the later 1730s (Lead 1977, 7). The furnace does not appear under either Leek or Horton in the lists of 1717, 1794 or 1750–88, or in Schubert (1957).

Madeley, Staffs. SJ 7744 [118]

A furnace at Madeley was in the hands of Walter Chetwynd in 1649; the previous tenants were William Yonge and Richard Foley. It was working in 1667 and was visited by Robert Plot in 1686, when it was being worked by William Chetwynd. During the 1690s Richard Chetwynd was supplying bar iron to Rugeley slitting-mill. In 1717 the furnace was said to have an output of 400 tons; it also appears in the locally compiled list of furnaces of 1735 and in a pamphlet advocating the building of the Trent & Mersey Canal in 1766.

Madeley is not in the 1794 or 1796 surveys and appears in the list of closures between 1750 and 1788, bracketed with Meir Heath and Lawton as being in the hands of 'Mr Cotton', with a note that it had been converted to a foundry. When this conversion took place is unclear. The works was leased in 1781 and again in 1795, although neither of these references explicitly mentions a furnace. Aikin (1795, 103) does refer to ironsmelting at Madeley, although his information, published in 1795, may have been collected a few years earlier. A lease of the site in 1795 from John Crewe to William Hill, Caleb Hill and John Stevenson (Cheshire RO, DCR 50/4/16) also includes a furnace, although it may not then have been in use. Lead noted a number of references to founders in Madeley parish register in the late eighteenth century and suggested that the furnace may have concentrated on castings; alternatively, these entries may reflect the conversion of the works to a foundry only (Lead 1977, 3, 7; Awty 1957, 73, 79, 99–100, 113, 116; Schubert 1957, 381).

Meir Heath, Staffs. SJ 9339 [127]

In the 1670s Richard Foley of Longton, son of Richard Foley of Stourbridge, was operating the furnace at Meir Heath. He died in 1678 and the works remained in the family; accounts survive for 1688–9, when it was part of the 'Moorland Works' operated by John Wheeler within the Foley organisation. The furnace was visited by Plot in the 1670s and in 1717 was said to have an output of 600 tons a year. The last lease of the works surviving among the muniments of the ground landlords, the Leveson-Gower family, dates from 1727 and was to Edward Hall and partners (Staffs. RO, D593/I/3/20). The Knights' accounts show that Warine Falkner supplied pig to the Stour Partnership from Meir Heath in 1728–9 and that Hall & Co. made similar sales in 1743–50 (Ince 1991, 117). In 1751 John Smith was referred to as the clerk at Meir Heath, which in 1754 was visited by Charles Wood, when it was producing cold-short pig with charcoal as fuel (Hyde 1973, 39). By 1763, however, the furnace was disused; it does not appear in the 1794 list and in the list of closures between 1750 and 1788 is bracketed with Madeley (Staffs.) and Lawton (Cheshire) as having been operated by 'Mr Cotton' but now converted to a corn-mill (Lead 1977, 1, 3, 5, 7; Awty 1957, 79, 114; Schubert 1957, 381).

Street, Cheshire SJ 8058 [118]

Little is known of this furnace, which lay just inside Cheshire south of Congleton. It may have been working in 1664, in the hands of William Fletcher of Makeney (Derbys.), but nothing certain is known until 1701–2, when it was converted into a plating-mill. The furnace was only two miles from the much bigger works at Lawton (q.v.) and may have been shut down to conserve local wood supplies (Lead 1977, 3; Awty 1957, 76, 79, 89; Schubert 1957, 388). Street does not survive as a place-name on the 1:50,000 map but the approximate site of the furnace is presumably indicated by the name Forge Farm on the A50 north of Alsager.

Oulton: see Vale Royal

Teanford, Staffs. SK 0040 [128]

This appears to be a short-lived, marginal site, a couple of miles south of Cheadle, which is referred to in the local list of furnaces compiled in 1735 (Lead 1977, 7). It is not in any of the nationally compiled lists, nor Schubert (1957). Nothing appears to be known of its history apart from the reference in 1735.

Vale Royal, Cheshire SJ 6369 [118]

It is not clear when this important furnace was built, or by whom. Awty suggests that it might have been the Foleys, if only because no-one else operating in the region in the late seventeenth century would have had the necessary capital. The furnace was unusual in being built a long way from its supplies of iron ore; in addition to local ironstone Vale Royal used Furness haematite. Accounts survive from the 1690s among the Foley papers and in 1696 Daniel Cotton became manager at Vale Royal on behalf of the family's Cheshire Works. In 1706 he was succeeded by Edward Hall. In 1716 Vale Royal was taken over by Abraham Darby and partners but for how long it remained in their hands is unclear; their agreement to operate the furnace was renewed in 1720. There is no reference to the import of haematite for the furnace in the Weaver Navigation books, which points possibly to closure before 1733 (Awty 1957, 79, 87–8, 93–4, 99; Schubert 1957, 390; Raistrick 1970, 42-3).

In 1717 Vale Royal had an output of 600 tons a year; it is not in the 1794 list and in the list of closures between 1750 and 1788 it appears as 'Oulton' (a nearby settlement), with the operator given as Kendall and the marginal comment 'down'. It does not seem possible to establish a precise date of closure.

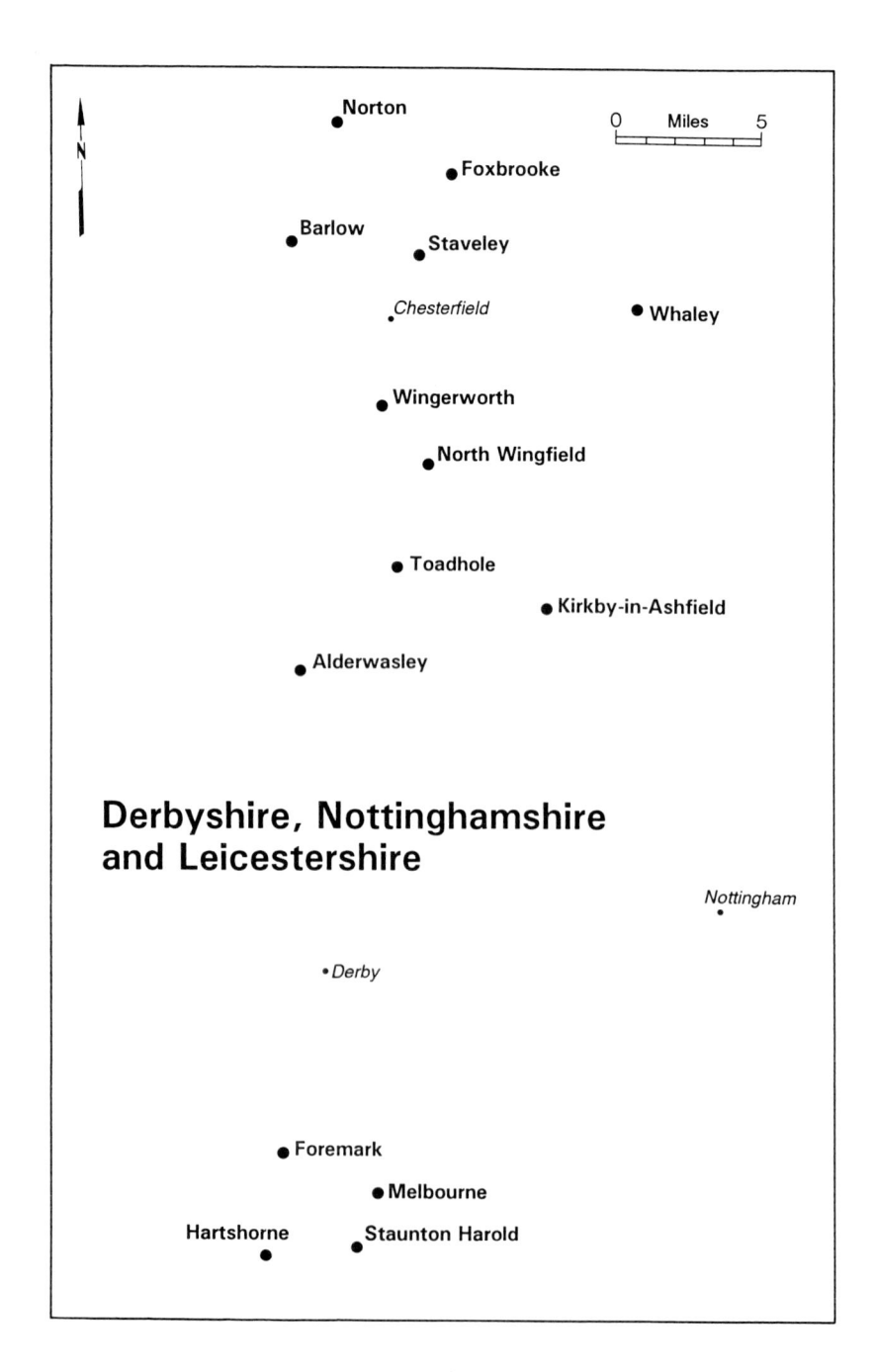

Derbyshire, Nottinghamshire and Leicestershire

7

Derbyshire, Nottinghamshire and Leicestershire

In addition to the sources cited in each entry in this chapter, a general survey of the charcoal iron industry in the East Midlands has recently been published (Riden 1991), which contains full archival references for statements that have not previously appeared in print.

Alderwasley, Derbys. SK 340524 [119]

About 1962 remains were discovered of a blast furnace at the engineering works of R. Johnson & Nephew on the west bank of the Derwent in the township of Alderwasley, about a mile north of Ambergate, and the site was included in the Historical Metallurgy Group's very first list of furnaces the following year. By 1965 all visible evidence had been destroyed (*Bull. Hist. Metallurgy Group*, Nos 1 and 4; Nixon 1969, 217–18).

No survey appears to have been made of the structure, although a few photographs have come to light, one of which has recently been published (Riden 1989, 176). This shows that the furnace was built against the natural bank that flanks the west side of Johnson's works and was thus some feet above the level of the Derwent. The bellows must therefore have been powered not by the river but by a tributary (Peatpits Brook) which runs down the steep hillside to the west and on which the modern OS map marks a small storage pond about half a mile upstream from the furnace (SK 520332). The photographs of the early 1960s show that about half the circumference of the lining and a good deal of the outer wall and rubble core of the furnace survived almost to full height. Notes on the back of one of the pictures record the stack angle above the boshes as 82° and the inward angle of the boshes as 70°. The furnace was said to measure about 9ft (2.7m.) across the top of the boshes and 4ft (1.2m.) at the throat; the overall height was estimated at 20ft (6.1m.) (Riden 1988, 80). After the furnace was demolished in 1964 a concrete floor was laid, sealing any surviving foundations, but nothing has since been erected on the site.

The investigators of the early 1960s were able to discover only one published reference to a furnace at Alderwasley, in a county directory of 1895. According to this, 'Near the works [of R. Johnson & Nephew] are the ruins of a blast furnace, erected in 1764, for the manufacture of iron for nails and sheets. Charcoal was exclusively used, and the ore was brought by packhorse'. At first sight, this statement seems improbable, since the date is so late for a new charcoal furnace, but it has recently been possible to reconstruct a fuller account of the Hurt family's involvement in the iron trade in mid-Derbyshire (Riden 1988, 1989), which demonstrates that the furnace at Alderwasley was built by Francis Hurt, almost certainly c. 1764, and did initially smelt with charcoal. About 1780 Hurt erected one of the first coke-fired furnaces in Derbyshire a few miles away at Morley Park. He then gave up smelting at Alderwasley, although the forge there continued to work in tandem with Morley Park until the furnaces were abandoned in the 1870s. Alderwasley remains in use as an engineering works today, while the substantially complete furnaces at Morley Park (SK 380492) form a scheduled monument currently being conserved by the Derbyshire Archaeological Society, which now owns the site.

What is perhaps most interesting about the Alderwasley furnace is that an analysis of slag recovered from the site in 1964 indicated that at some date during its fairly short working life coal or coke (or both) had been used for smelting there. Indeed, Alderwasley may have been the last charcoal furnace to be built anywhere in England in the eighteenth century or the first coke furnace in the East Midlands or both. It is therefore particularly regrettable that all trace of the structure itself has disappeared.

Barlow, Derbys. SK 351753 [117]

A furnace was established on Barlow Brook by the sixth Earl of Shrewsbury in 1605–6, with an associated forge. The latter had become a corn-mill by the 1650s but the furnace survived and in 1693 was leased for 11 years to John Jennens of Erdington Hall (Warwicks.). This seems to be the last reference to the site, which does not appear in any of the eighteenth-century lists (Schubert 1957, 367; Hopkinson 1961, 123–4, 132; Riden 1991, 67, 69). The site is indicated today by the name Furnace Lane, where structural remains were reported in about 1950 which have since been cleared away (Mott 1949–51, 235; Nixon 1969, 227).

Foremark, Derbys. SK 330270 or 331235? [128]

Farey (1811, 396, 401) includes 'Foremark, N.W. of the Park' in his normally reliable list of former bloomery and charcoal furnace sites. No evidence has yet been found for the history of this site, although it may be significant that the major owner in Foremark was the Burdett family and that in 1689 Elizabeth Burdett married Charles Jennens of Erdington, whose family had the nearby Hartshorne furnace around the same date. Cranstone (1985, 276–8) suggested two possible sites for the furnace which would fit Farey's directions and commonsense topographical deduction (SK 330270 or SK 331235), either on the parish boundary with Repton, north-west of the park near Foremark village, or on the same boundary north-west of Foremark Park farm. The latter, which is now beneath Foremark Reservoir, seems more likely. The field-name 'Furnace Close' at SK 326258 appears to be a red-herring, since Cranstone considers the site unsuitable for any water-powered industry and there is no surface evidence.

Foxbrooke, Derbys. SK 428772 [120]

The furnace at Foxbrooke near Eckington was built in 1652 by George Sitwell of Renishaw Hall in the same parish and became the nucleus of his extensive ironmaking enterprise in Derbyshire and Nottinghamshire. After the family withdrew from direct involvement in the industry in the 1690s, Foxbrooke was leased to a partnership centred on the Foley family of Herefordshire and managed by John Wheeler of Woollaston Hall, Stourbridge. After a few years the Foleys gave up their Derbyshire and Nottinghamshire works, which passed to the Spencer partnership, whose other interests lay in south Yorkshire. Foxbrooke was abandoned in 1749 and converted to a sickle-mill. The site is readily identifiable today at Old Furnace Wood, halfway between Slitting Mill Farm (itself the site of another Sitwell works) and a large dam on Fox Brook in Foxstone Wood, which powered the furnace (Riden 1985; 1991, 71, 73–4). In 1717 the furnace was said to be producing 150 tons of pig a year.

Hartshorne, Derbys. SK 326213 [128]

This is another site on the Ashby coalfield where Farey (1811, 401) spoke of an old charcoal furnace. Leases of 1699 and 1702 indicate that it was then at work and was in the hands of John Jennens of Erdington (Cranstone 1985, 28). It was described as disused by William Woolley in about 1712 (Glover and Riden 1981, 154) but appears (as 'Hawthorn') in the list of closures between 1750 and 1788, bracketed with Kirkby (q.v.) and said to have been operated by the Mathers, who succeeded Jennens at Kirkby and elsewhere. Farey, whose dates are usually reliable, said that it had been abandoned about 50 years previously (i.e. about 1760), which seems reasonable. Spavold (1984, 109–11) has located the site, which was later used by a screw-forge.

Kirkby-in-Ashfield, Notts.   SK 4855 [120]

Kirkby, the only charcoal blast furnace in Nottinghamshire, was built under a lease of 1673 from the Duke of Newcastle to Humphrey Jennens of Erdington Hall (Notts. Archives, DDP 15/61; Schubert 1957, 379; Hopkinson 1961, 124). In 1717 output was said to be 200 tons p.a. The furnace later passed into the hands of the Spencer partnership, whose surviving accounts show that Kirkby was out of blast throughout the period 1750–65 (Hopkinson 1961, 133). The site is included in the list of closures between 1750 and 1788, bracketed with what appears to be a reference to Hartshorne in south Derbyshire (q.v.), with the occupier's name for both given as 'Mather', i.e. Walter Mather, whose family's connection with the iron industry seems to have begun at Bulwell forge in the early eighteenth century. In 1783 Mather took a lease of the furnace at Staveley (q.v.), where he built a new works, and then gave up his remaining charcoal ironmaking interests, presumably including Kirkby furnace, although an exact date of abandonment has not been traced (Riden 1991, 73, 78). Johnson (1960, 46) located the probable site of the furnace near Meadow Farm.

Melbourne, Derbys.   SK 379239 [128]

Burdett's map of Derbyshire (1762–7) marks a furnace south of Melbourne on a stream called New Brook, which flows south into the Trent. This can be identified with a furnace described as 'down' in the list of closures between 1750 and 1788 and said to have been worked by Messrs Lloyd. The site was located by W.H. Bailey in the early 1960s and excavated by him before the area was flooded beneath Staunton Harold Reservoir (*Bull. Historical Metallurgy Group*, No 3, 1964; Smith 1965, 123; Nixon 1969, 55, 268; no further report on these excavations has appeared, although cf. Blick

1984, 44–5). Bailey believed that the furnace was established about 1720, although the earliest firm evidence seems to be its appearance on an estate map of 1735. The furnace was leased by the Earl of Huntingdon to the Lloyds of Birmingham in 1758 as their only venture into smelting and worked in conjunction with a forge at Burton-on-Trent (the lease is now Leics. RO, DE 4210/2/8 and the estate map DE 362/1; cf. Lloyd 1975, 146–8, 176, 280–1). The furnace was surrendered by the Lloyds in 1772 and early the following year the bellows were removed and taken to Wingerworth furnace, presumably to be reused there (Riden 1991, 73n.).

North Wingfield, Derbys. SK 409636 [120]

A plan of 1621 of the Park Hall estate in the parish of North Wingfield drawn by William Senior for Sir Francis Leeke of Sutton marks a blast furnace and associated dams and goits (but not a forge) on the Rother between Danesmoor and Park House Green. This is the earliest evidence for a furnace on this site, which was presumably built some time in the previous twenty years. It was operated by George Sitwell of Renishaw in the 1660s but has no later history; Hopkinson offers no evidence for a date of closure as late as 1700 (Schubert 1957, 382; Hopkinson 1961, 132; Riden 1985; 1991, 68, 69).

By the time of Burdett's map of 1762–7 the site had become a corn-mill. Its connection with the iron industry was first noticed archaeologically by George Griffin (1918, 202–6), who identified the site of Park Hall, the medieval mansion of the Deincourt family, remains of which were encountered during the building of Park House Colliery in 1867. During the same operation Griffin noticed 'a stratum of black slags, the residue of ancient iron smelting, operations, in which the fuel used was charcoal or wood ... This stratum of slag was buried to a depth of four feet by the levelling down of the debris resulting from the final demolition of the ruins [of the mansion]'. The suggested stratigraphy is probably confused, since slag from the furnace must have overlain remains of the hall, unless there was also slag from an earlier bloomery on the site, but the features noticed by Griffin correspond with the field-name Cinderhill on the North Wingfield and Pilsley tithe maps of the 1840s. Much of the area around the furnace has been completely altered by opencast coalworking but the site of the furnace itself, in a field next to Park House Farm which displays slight surface irregularities, does not appear to have been affected by subsequent building, mining or other activities.

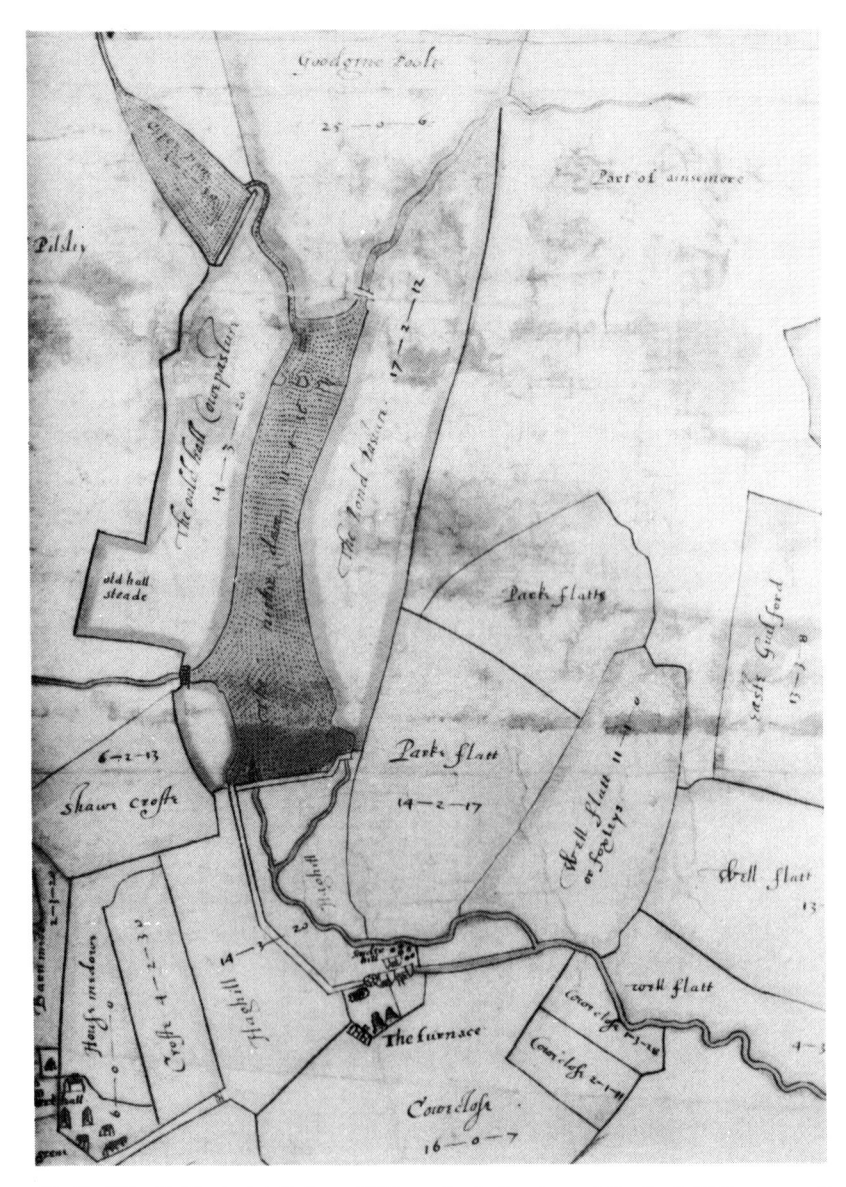

North Wingfield. A detail from William Senior's plan of the Park Hall estate of 1621 showing Sir Francis Leeke's furnace a short distance west of the mansion itself. The plan is drawn with south at the top. *(William Salt Library, Stafford)*

Norton, Derbys. SK 3582 [110]

George Sitwell's letterbook includes references to the Bullock family of Norton which imply they were then operating either a furnace, forge or both on their estate there; this is also borne out by a reference to ironworks at Norton among the papers of the Committee for Compounding (Riden 1985). In 1669 Lionel Copley entered into an agreement with Gervase Nevile to take over Norton forge and furnace, subject to confirmation from Mrs Bullock, widow of William Bullock (died 1666), who was evidently operating the works in Sitwell's day (ex inf. John Goodchild). The furnace seems to have no later history; Copley died in 1675.

Staunton Harold, Derbys. *c.* SK 377221 [128]

A furnace here is apparently first mentioned in a Chancery action of 1614 (PRO, C 2/A3/31, ex inf. B.G. Awty per D. Cranstone), which refers to a lease of ironworks at Staunton by Sir George Shirley. Since the lessee was described as an 'ironfounder' the works evidently included a blast furnace. In 1624 the furnace was leased to John Wenham of Battle (Sussex), the only instance yet noted of Wealden investment in the East Midlands industry. The only other reference to the site appears to be in the list of furnaces closed between 1750 and 1788, in which it appears as 'Stone Herald' with no details of occupier. The furnace may have been revived in the late seventeenth or eighteenth century, although it is not included in the 1717 list, or it may have been long derelict when it was listed in 1788. Cranstone (1985, 29–30) reported the discovery of slag on the bank of Staunton Harold Reservoir, which probably corresponds with similar finds by W.H. Bailey during his excavation at Melbourne (q.v.), which he described as being at Calke. The furnace site itself appears to be beneath the reservoir. Remains reported in a vague fashion by the local geologist Edward Mammatt in the 1830s (Smith 1965, 123) may belong to the Staunton Harold furnace or to medieval ironworking (cf. Cranstone, loc.cit.).

Staveley, Derbys. SK 417751 [120]

The furnace and forge about a mile west of the town of Staveley in the Rother valley are apparently first mentioned in a survey of the manor of Staveley in 1639 (Hopkinson 1951–7b, 106), although they may have been established by the Frecheville family of Staveley Hall at any date in the previous twenty years. In the 1650s and 1660s the works were leased by William Clayton of Whitwell and George Sitwell of Renishaw, descendants of the latter remaining tenants until the 1690s (Riden 1985, x). After the Sitwells' retirement from direct involvement in the industry Staveley was leased to John Jennens of Erdington (Hopkinson 1961, 134), but at the beginning of the eighteenth century passed to the south Yorkshire group of ironmasters centred on William Spencer of Cannon Hall. The Notts. & Derby group belonging to this partnership included, in the early part of the century, Staveley and Foxbrooke furnaces (q.v.), Staveley and Carburton forges and Renishaw slitting-mill. The partnership was reconstructed at various dates and in 1765, when the Spencers retired from the industry, was taken over by a Sheffield group (Hopkinson 1961, 134–6, 148; 1951–7a; Raistrick & Allen 1939). In 1783 the lease from the Duke of Devonshire came to an end and, after some negotiations, was not renewed to the previous tenants. Instead, a new lease was made to Walter Mather, who rebuilt the works on modern lines with a coke furnace (Chapman 1981, 12–17; Riden 1991, 74–8). All trace of early remains at Staveley has long been swept away by the massive expansion of the iron, engineering and chemical business on the site since the 1780s.

Toadhole, Derbys. SK 390568 [119]

The furnace here, whose site is still indicated by the name Toadhole Furnace on a tributary of the River Amber in Shirland parish, was probably built shortly after 1609, when a document discusses the prospects for its construction. It was operated initially by the Earls of Shrewsbury in conjunction with a forge at Crich Chase in the Derwent valley but, apart from stray references in the 1620s and 1640s, little is known of its early history (Riden 1991, 67, 69). A map of 1684 marks 'A Furnis of Toadhole', presumably meaning an ironworks rather than merely a place, but this document cannot at present be located. Toadhole does not appear in the 1717 list but is in that of closures between 1750 and 1788, lacking an occupier's name but with the marginal note 'Tan yard'. The site has now been cleared but slag and other debris survived there within living memory, while the names Upper and Lower Delves farms higher up Alfreton Brook point to ironstone mining in the area. Highway surveyors' accounts of 1751–2 for Shirland speak of loads of stone from the 'Furnace pitts' and of repairing 'Furnis Bridge' (Turbutt 1977, 63–4, 154).

Whaley, Derbys. SK 5270 [120]

A blast furnace was established on the Earl of Kingston's estate at Whaley in 1617 and operated in conjunction with a forge further east at Cuckney, The two works were built by Martin Ash, who soon became insolvent, and both furnace and forge reverted to Kingston, who still had them in hand in 1632 (ex inf. P.W. King from PRO, C 2/C6/37; Riden 1991, 68). The furnace is mentioned in a survey of Scarsdale Hundred of 1652 but then seems to be undocumented until the list of 1717, in which it appears as 'Wanley', with an output of 300 tons a year. Farey (1811, 395) suggested that Whaley closed about 1777. The site is marked and named on Burdett's map of Derbyshire (1762-7), although, according to Hopkinson (1961, 133), the Spencer partnership's records show that Whaley was only in blast twice in the period 1750-65.

Although Farey speaks of the furnace being at 'Whalley in Bolsover', it was in fact in Scarcliffe parish. The area around the furnace was completely altered in the nineteenth century by railway building and it is not clear whether the site was re-used, although unlike some others in the district it did not become a cotton-mill. The ground landlords of the furnace in the late seventeenth century, as successors to the Pierreponts, were the Bathurst family of Cirencester, whose muniments are now in the Gloucs. RO (D2525). Lane (1973) printed part of the record office schedule and noted the field-name 'Furnace Fall' in Scarcliffe Park, immediately NW of the furnace site. Lane's fieldwork, while revealing evidence for Roman leadsmelting, did not uncover any trace of early ironworking.

Wingerworth, Derbys. SK 382661 [119]

According to Farey (1811, 395) a charcoal furnace at Wingerworth which closed in 1784 had been in existence for more than 180 years. Because of the loss of most of the muniments of the ground landlords, the Hunlokes of Wingerworth Hall, it is difficult to confirm that the site was among the earliest furnaces in the region, although deeds abstracted in the eighteenth century show that a bloomery was the first property the family acquired in the parish in 1547, a generation before they bought the manor and most of the land. The deeds do not mention a furnace but it seems reasonable that the Hunlokes should have experimented at an early date with the new technique, added to which Farey was normally well informed in matters to do with the iron industry (Riden 1991, 67-8).

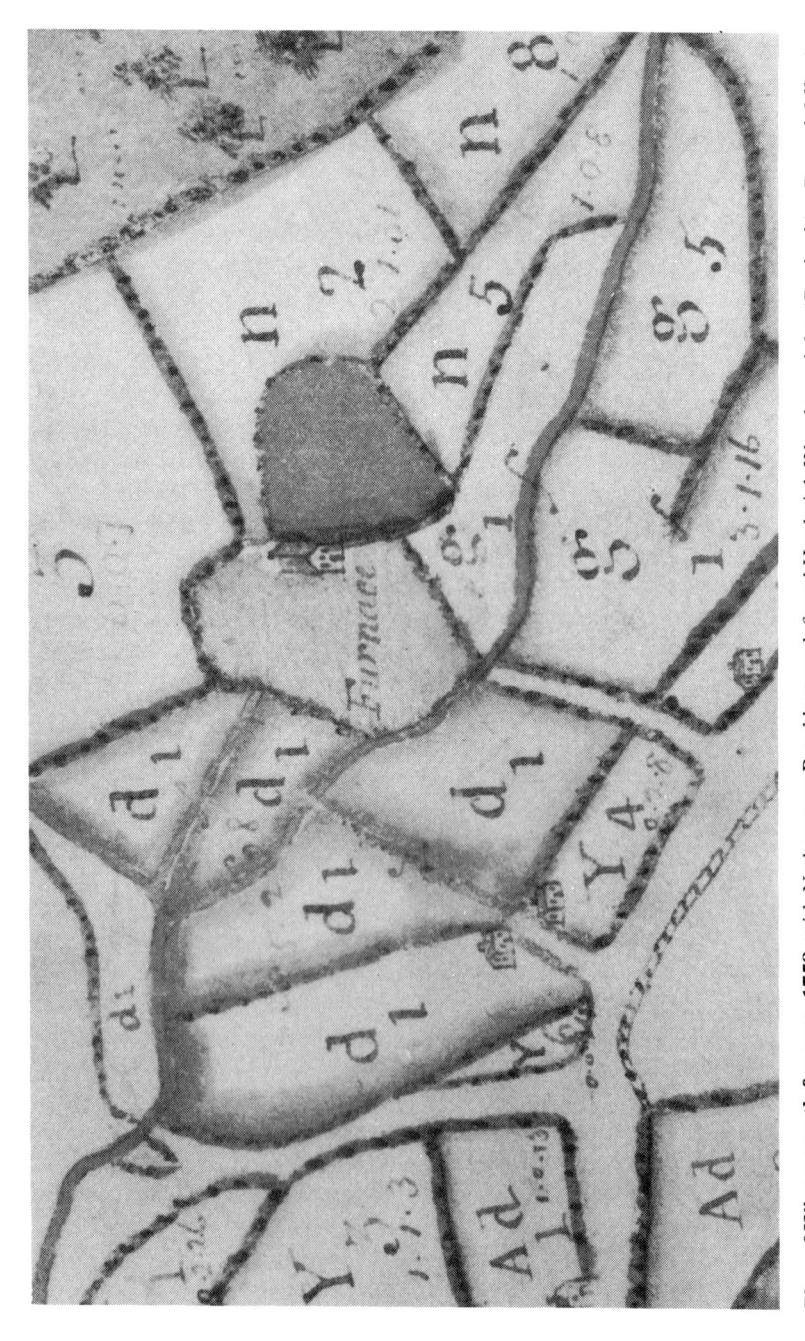

Plan of Wingerworth furnace, 1758, with Nethermoor Road bottom left and Hardwick Wood top right. *(Derbyshire Record Office)*

The furnace was still in use, together with a forge lower down Tricket Brook, in the 1650s, when it was mentioned in an Exchequer suit, and from 1681 a series of leases survives, covering the last century of its history (Riden 1991, 68–9, 71, 73, 74). In that year the furnace was leased to a Birmingham ironmonger Thomas Pemberton, who succeeded a local lead merchant named Thomas Bretland, who was possibly the Hunlokes' first tenant, since the furnace had been in hand 30 years before. In 1702 John Jennens of Erdington took over Wingerworth, which in 1710 passed to a third Birmingham ironmonger, Riland Vaughton, who was followed in 1725 by a syndicate of local men headed by the Chesterfield lead merchant William Soresby. In 1741 a new lease was granted to Walter Mather of Bulwell and John Mander & Co., another Birmingham firm; ten years later Mather alone became tenant, but was joined in 1758 by the landlord's brother James Hunloke. Mather continued to work the furnace until 1777; account heads have been ruled in a surviving ledgers for 1778 but not used, although there is no explicit statement that the furnace was blown out.

The Hunlokes made a new lease of their ironworks in 1781 to the Broseley (Shropshire) iron merchant George Matthews and a York land surveyor named Joseph Butler. This included the charcoal furnace but Matthews & Butler proceeded to build a new coke-fired works a short distance away. The lease of 1781 was advertised for sale three years later, when the charcoal furnace was said to be still in use; since Farey's date of closure of 1784 was probably supplied by Butler's son (also called Joseph), who later developed a considerable ironmaking business in the Rother valley (Riden 1984), the furnace almost certainly shut in the year in which the partnership was reorganised.

The charcoal furnace is marked on an estate map of 1758; the field in question remains unbuilt-on and contains indeterminate earthworks. The site of the coke furnace of 1781 (SK 384661) was landscaped in 1973, when the base of a furnace stack was uncovered (Riden 1973).

N

• Seacroft

. Leeds

Bank
• Bretton

• Barnby
. Barnsley

• Rockley

• Chapel

• Masbrough
Wadsley
•

• Sheffield

Yorkshire

0   Miles   5

8

Yorkshire

The history of charcoal ironsmelting in Yorkshire in this period is in general well worked out, thanks largely to the survival of the accounts and papers of the Spencer family of Cannon Hall (now Sheffield Archives, Spencer Stanhope MSS), who were at the centre of a group which controlled most of the industry in the late seventeenth century and first half of the eighteenth, and their use by a succession of writers (Raistrick and Allen 1939, Raistrick 1939, Awty 1957, Hopkinson 1961). The changing composition of these partnerships cannot be dealt with in a work of this kind and throughout this chapter the term 'Spencer family partnership' has been used as shorthand for a group which included several other, related families.

Bank, Yorks. SE 2613 [110]

Two furnaces, known as Upper and Nether Bank, were operated by the Spencer partnership under an agreement of 1696 until the mid-eighteenth century (Schubert 1957, 366). The first furnace on the site appears to have been established about 1640, replacing an earlier bloomery (Mott 1971, 68–9). Raistrick and Allen (1939, 168, 171–2) estimated that the Upper Furnace produced an average of 400–450 tons of pig annually, the Lower Furnace 150 tons. In the 1717 list an entry appears for 'Banks' (possibly meaning the two together) with an output of 400 tons. The Bank furnaces were worked in partnership with William Cotton of Haigh Hall, who acted as manager, and supplied pig to Knottingley forge.

In the list of closures between 1750 and 1788 'Bank near Barnsley' appears with 'Cookburn' as occupier and the marginal comment 'down'. John Spencer of Cannon Hall noted in his diary in August 1774 that Bank was then shut down for lack of charcoal (Hopkinson 1961, 133); it is not clear whether it ever worked after this date.

A furnace named 'Wortley' also appears in the list of closures between 1750 and 1788, which was said to be 'down' but no details are given of the occupier. Under the same name in the 1794 list is an entry for a works owned by the 'Earl of Bute's Lady' (i.e. Mary, daughter of Edward Wortley

Montague of Wortley, wife of John, 3rd Earl of Bute, created Baroness Mount Stuart *sua jure* 1761, who died in 1794), occupied by J. Cockshutt, with two charcoal furnaces, four fineries, two chaferies and a rolling- and slitting-mill. The words 'Wire Mills' have been written across the columns ruled for the method of blowing and date of building, while a marginal comment says 'Tin Mill'. The forge and mill plant listed is the complex known as Wortley Forges, part of which, the Top Forge (SK 295999) is now a scheduled monument; the history of the site has been worked out by R.A. Mott (1971). There were, however, no furnaces at Wortley itself, which worked in tandem with those at Bank, which are presumably the two referred to in the 1794 list. Bank furnaces and Wortley forges were leased together from the Spencers' time onwards; after they had retired from business (apparently in 1765) the lease was taken by John Cockshutt, who was succeeded by his son of the same name in 1789 and it is therefore he who appears in the 1794 list.

Forging continued at Wortley into the twentieth century, but the charcoal furnaces evidently shut before 1796, since they do not appear in the list of that year or any later returns. Possibly their appearance in both the 1794 list of furnaces in use and that of closures during the previous 30 years points to intermittent use after the Spencers had given up the lease. The identity of 'Mr Cookburn' remains unclear, unless it is an error for Cockshutt (which seems rather unlikely).

The approximate site of the two Bank furnaces is indicated on the modern map by Bank Wood, between Emley and Bretton Hall, where remains of bell-pits from which ironstone was mined can be seen (Hey 1986, 83, 224, 225).

Barnby, Yorks. SE 2908 [110]

The furnace here was the nucleus of the Spencer family's activities in south Yorkshire and was already in operation when Randolf Spencer of Criggion, Montgomeryshire, sent his son John to act as clerk to his relation Major Walter Spencer at Barnby in 1650. The furnace initially supplied a forge at Kirkstall, which in 1658 was leased by the owner, Viscount Savile, to John Banckes, Russell Allsop and William Fownes, all London merchants, a lease that included Barnby furnace (cf. Sheffield Archives, Sp.St. 60495). In 1675 Banckes and his co-lessees released Kirkstall and Barnby to Thomas Dickin and William Cotton (the latter a nephew of Fownes). The following year Dickin and Cotton brought John Spencer, son of John Spencer of Criggion (died 1658), into partnership at Kirkstall and Barnby. The furnace appears to have succeeded a bloomery sometime between 1635 (when the bloomery is mentioned) and 1658. In 1696 Cotton, Dickin and Spencer brought a number of other south Yorkshire works into a much larger partnership that continued to operate, with changes in personnel, until the death of William Spencer in 1762 (Raistrick and Allen 1939, 168–71; Schubert 1957, 367).

In 1717 Barnby was said to produce 300 tons of pig annually; Raistrick and Allen (1939, 172) suggested from the surviving accounts a figure of 400 tons. The list of furnaces closed between 1750 and 1788 includes Barnby, said to have been operated by the 'Colnbrook Co.' and to be 'down'; in August 1774 John Spencer recorded that it was shut for want of charcoal (Hopkinson 1961, 133) and it may not have worked again.

Bretton, Yorks. SE 2812 [110]

This furnace does not appear in the 1717 list but was probably built shortly afterwards, since under an agreement of 1720 Edward Hall of Cranage, Thomas Cotton of Doddlespool and Samuel Shore of Kilnhurst were to operate Bretton for 21 years as equal partners (Awty 1957, 100). It continued to be operated by the partnership centred on the Spencer family until the middle of the eighteenth century, with an output estimated from surviving accounts of about 450 tons a year (Raistrick and Allen 1939, 169, 172; Schubert 1957, 369; Hopkinson 1961, 135, 139). It was then worked by Thomas Cotton, son of William Westby Cotton (died 1749), together with a foundry. In 1774 John Spencer noted that only Bretton was still in blast in the Barnsley area, Bank and Barnby being shut for want of fuel (Hopkinson 1961, 133). It appears in the 1794 list as 'Britton', five miles from 'Wandfield' (i.e. Wakefield), owned by Colonel Britton, occupied by Messrs Cook & Cockshutt, with a single charcoal furnace, although it may have shut down temporarily just before this date, probably when Cotton gave up the lease. In 1796 the furnace was said to be making 250 tons a year; in 1805, still charcoal-fired, it had a similar output, but shut permanently the following year (ex inf. John Goodchild).

Chapel, Yorks. SK 3596 [110]

A furnace at Chapeltown, at the head of the Blackburn valley, is apparently first referred to in 1628 and appears in a survey of the manor of Sheffield of 1637 (Schubert 1957, 370; Hopkinson 1961, 124). In 1639 Chapel furnace and other ironworks in south Yorkshire were leased by the Earl of Arundel to Lionel Copley and others. Chapel furnace and Kimberworth furnace subsequently passed to Francis Nevile of Chevet, who in 1652 sub-let both works to Copley (ex inf. John Goodchild; cf. Hey 1971–7, 252–3). After Copley's death in 1675, Chapel and several nearby forges came into the hands of William Simpson, an attorney. In 1700 the partnership controlling these works included Simpson's son John; Francis Barlow of Middlethorp (Yorks.), a collateral descendant of Francis Barlow of Sheffield, who had been one of William Simpson's descendants; and Dennis Heyford of Staveley (Derbys.), a descendant of the steward of Sir Francis Rockley, a Puritan, who had used his influence with the Committee for Compounding to secure outlawry for his master and the ironworks at Rockley for himself.

The partnership operating Chapel, Rockley and Bank furnaces, plus a number of forges, known together as the Duke of Norfolk's works, was reconstructed in 1727 when John Spencer of Cannon Hall and his associates were brought in. Another major restructuring took place in 1765 following the death of John Fell and William Spencer. The assets were now acquired by John Cockshutt of Wortley forge; John Travers Younge, a Sheffield merchant; Richard Swallow, who had acted as clerk to the partnership after Fell's death; and John Clay (Hopkinson 1961, 134–6; Raistrick and Allen 1939, 168–72). The new company operated on a reduced scale in both Derbyshire and Yorkshire, with furnaces at Staveley (q.v.) and Chapel. The Derbyshire works were given up with the expiry of the Staveley lease in 1783; when Chapel closed is not clear, although it appears in the list of charcoal furnaces abandoned between 1750 and 1788, with no additional information. There was an unsuccessful attempt to use coke at Chapel in 1759 and the furnace was laid off in 1762. The 1794 list has an entry for the new coke furnace built on a nearby site by Richard Swallow, apparently before 1779 (Hey 1971–7, 255).

In 1717 Chapel was said to make 200 tons a year; Raistrick and Allen's estimate (1939, 172), based on surviving accounts, was 450 tons. The site has been located by Hey (1971–7, 252–4) but no surface remains survive.

Masbrough, Yorks.   SK 4192 [111]

The 1794 list includes the Walkers' works near Rotherham, with the owner listed as Lord Effingham and the plant then consisting of two furnaces fired with coke and one with charcoal, a blowing-engine, three fineries, a chafery, a balling furnace, and a rolling- and slitting-mill. A date of building of 1765 is given for the furnaces, without specifying to which of the three this relates (cf. Schubert 1957, 381). These details correspond with the company's own records (John 1951, ii, 5), which show that a charcoal furnace was built about 1760 and remained in use until at least 1791. John suggested that the first coke furnace was not built until the late 1760s but possibly the date given in the list is correct. The second coke furnace was added in 1779. In the lists of 1796 and later the works at Masbrough are listed with three large coke-fired furnaces only.

Rockley, Yorks.   SE 338022 [110]

The furnace here was built by Lionel Copley in 1652, succeeding a bloomery forge half a mile away to the east (Crossley and Ashurst 1968). After Copley's death in 1675 Rockley passed into the hands of Dennis Heyford, who in 1696 became involved in the partnership centred on the Spencer family, in whose hands the furnace remained until the mid-eighteenth century (Crossley 1980, 445). In 1726 William Westby Cotton obtained a sixteen-year lease of the works (Awty 1957, 84, 100; Raistrick and Allen 1939, 168–72; Schubert 1957, 385; Hopkinson 1961, 134; cf. Chapel). Rockley is probably represented by 'Horchley' in the 1717 list, when it was said to have an output of 400 tons, about the same as that estimated by Raistrick and Allen from the Spencer accounts.

The list of furnaces closed between 1750 and 1788 includes one at Stainborough, which is presumably Rockley; no occupier is named and the marginal comment is simply 'down'. C.R. Andrews (1956) related a local tradition that the furnace was in use during the Napoleonic Wars, which can perhaps be accepted, given the discovery during excavations in the 1970s of a scatter of coke in the stocking areas on the charging-bank and a casting-pit in which ordnance may have been produced. On the other hand, since the storage ponds which served the furnace were probably drained before 1813, it seems unlikely that Rockley operated far into the nineteenth century (Crossley 1980, 445).

Rockley. The abutment of the charging-bridge can be seen on the right beyond the furnace.

Access to the site (a scheduled monument) is by a short, clearly signed public footpath off the minor road between Birdwell and Stainborough, not far from Junction 36 on the M1, which passes very close to the furnace. The principal feature at Rockley is the furnace itself, which stands to a height of 6m., with probably no more than 1m. lost from the top of the structure. The furnace was charged from a bridge extending from the stone-faced bank to the south and was tapped from the arch on the west side. The blowing-house lay on the north, with the wheel-pit at its eastern end; the purpose of the third archway into the furnace on the east side of the structure is not clear. The head-race ran to the wheel-pit via a conduit which passed to the east of the charging bank (cf. Crossley 1980 for a plan and fuller description).

Seacroft, Yorks.   SE 3536 [104]

A furnace at Seacroft, four miles NE of Leeds, was operated by the Spencer partnership between 1696 and the mid-eighteenth century, although it does not appear in the 1717 list (Schubert 1957, 387). Raistrick and Allen (1939, 169, 172) estimated average output at 300 tons from the Spencer accounts. In the list of closures between 1750 and 1788 Seacroft appears in the list headed 'Coak furnace' appended to the list of charcoal furnaces (in which it does not appear), which possibly suggests that after the Spencer partnership gave up the works an attempt was made to use coke there. The position is hardly clarified by an entry for Seacroft in the 1794 list, which fails to name either owner or occupier, and has the letter 'D' (for 'Down'?) in the column in which charcoal furnaces are listed, with no other plant. The furnace was said to be blown by water and to have been built in 1780.

Wadsley, Yorks.   SK 3290 [110]

According to Hopkinson (1961, 124) a furnace was built in Sheffield Park a few years after a survey of the manor of Sheffield was taken in 1637 and was later leased to Lionel Copley; according to Schubert (1957, 390), the furnace was in existence by 1585 and was leased to Copley in 1672. The furnace is also apparently mentioned in a survey of 1683 but does not appear in any of the eighteenth-century lists.

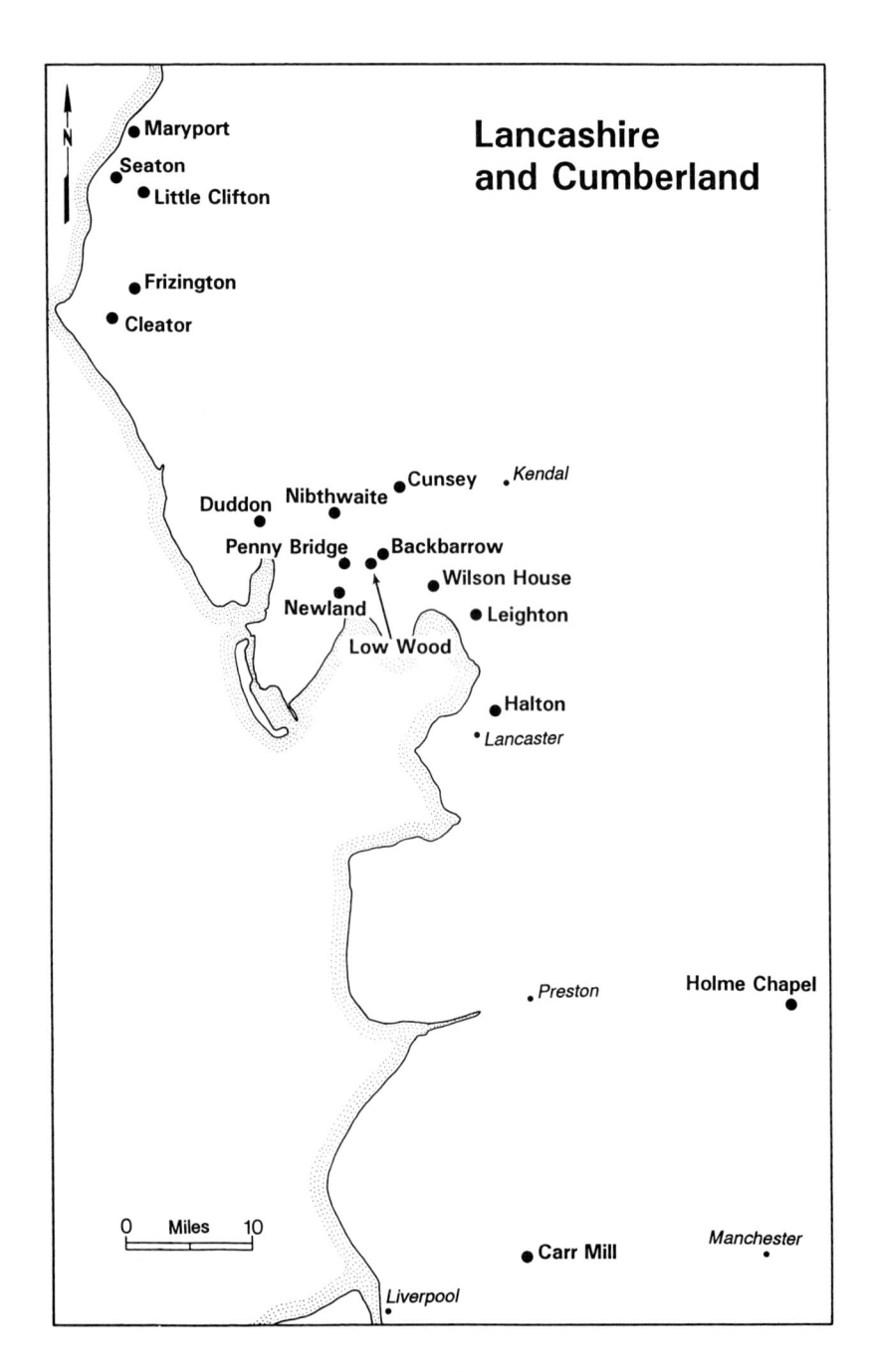

Lancashire and Cumberland

- Maryport
- Seaton
- Little Clifton
- Frizington
- Cleator
- Duddon
- Nibthwaite
- Cunsey
- Kendal
- Penny Bridge
- Backbarrow
- Wilson House
- Newland
- Leighton
- Low Wood
- Halton
- Lancaster
- Preston
- Holme Chapel
- Liverpool
- Carr Mill
- Manchester

0 Miles 10

9

Lancashire and Cumberland

This chapter includes the furnaces in Furness and Cumberland, together with two sites elsewhere in Lancashire. This region can boast not only the longest surviving charcoal furnaces in Great Britain, including one that continued into the present century, but also several early attempts at coke smelting. It should perhaps be added that no reference has been included here to M.J. Galgano's claim (1976, 212–13) that a blast furnace was built by Sir Thomas Preston of the Manor as early as 1664, since C.B. Phillips (1977, 34) has shown that the documents cited in fact relate to limekilns.

Because of the late survival of charcoal smelting in Furness, the region today has the best concentration anywhere in Britain of well preserved sites worth visiting, although all are on private property, whose owners should be approached before entering. Two points may be noted concerning the map references in this section. Firstly, most of the sites listed as being shown on Landranger 96 can also be found on sheet 97, while for serious exploration of the Lake District, the set of four 1:25,000 Outdoor Leisure maps (Nos 4, 5, 6 and 7) give a much fuller picture than the 1:50,000 sheets.

I am much indebted to David Cranstone, Andrew Lowe and Colin Phillips for help with this chapter.

Backbarrow, Lancs. SD 355846 [196]

The furnace at Backbarrow was built in 1711 and blown in for the first time in June 1712. It was operated by a company made up of William Rawlinson of Force Forge, John Machell of Backbarrow Forge, Stephen Crossfield of Plumpton and John Oliphant of Penrith. The original furnace was rebuilt in 1770 but coke did not replace charcoal as fuel until 1921. In the 1794 list the owner's name is given as John Machell and the occupier as 'B. Barrow Co.'; there was one charcoal furnace (wrongly said to date from 1705), a finery and a chafery (Fell 1908, 208–9, Barnes 1951, 82–4; Marshall and Davies-Shiel 1969, 37, 43–5, 223). James and later John Machell sold pig from Backbarrow to the Knights' Stour Partnership forges for most of the period 1747–99 (Ince 1991, 117).

In 1796 Backbarrow was said to have an annual output of 700 tons, probably meaning 14 tons a week for however long the furnace was in use each year. It reappears in the lists of 1806 and 1825, and then in *Mineral Statistics* from 1854 onwards. In 1806 an output of 446 tons was returned, which is sufficiently precise to suggest that it reflects actual production, whereas the 1,200 tons p.a. printed by Truran (1855) is probably an inspired guess.

The works finally closed in 1966 and has since remained derelict, although it has been accorded protection as a scheduled monument. On the site today, accessible from the minor road through Backbarrow village on the opposite bank of the River Leven from the main A590, the much rebuilt blast furnace can still be seen, together with some ancillary buildings. On the other side of the road through the village a track leads up to a large charcoal store standing alongside the former Furness Railway branch from Ulverston to Lakeside, which served the ironworks in the nineteenth century; the line itself here survives as the Lakeside & Haverthwaite Railway.

Carr Mill, Lancs. SJ 5297 [108]

In 1720 Edward Hall secured a 32-year lease of Carr Mill, in Ashton-in-Makerfield, from William Gerard of Garswood, Lancs., with licence to build a furnace and forge, rebuild the dam-head and create new mill-ponds. Considerable quantities of Staffordshire ironstone were shipped down the Weaver in the 1730s and 1740s and it is likely that much of this was destined for Carr, especially as shipments cease in 1751 on the expiry of the lease. There were also local purchases of charcoal. The lease was not renewed after this date and the furnace probably shut around the middle of the century (Awty 1957, 101, 109, 112). In 1749–50 and again in 1753–5 Hall & Co. sold pig from Carr to the Knights' Stour forges (Ince 1991, 117).

In the list of closures between 1750 and 1788 Carr Mill was said now to be a corn-mill occupied by one Rigley. The site reappears, however, as 'Carmill Wigan' in the list of coke furnaces appended to this list, which can apparently be explained by a lease of 1759 of Carr furnace to Samuel Johnson of Liverpool and two partners, George Perry and John Gosling, both described as ironmongers of Coalbrookdale (Awty 1957, 115). This appears to mark the beginning of coke-smelting in Lancashire but was evidently an abortive venture which closed before 1788. Carr Mill Dam is still a prominent landmark on the modern map, indicating the approximate site of the furnace.

Cleator, Cumberland   NY 014131 [89]

H.A. Fletcher (1881, 8–9; cf. Schubert 1957, 371) described remains that then survived at a corn-mill at Cleator, near Ehen Hall, of two sides of a blast furnace with a tymp-arch of about 10ft (3.0m.) span at its widest point. A slag heap was found by excavation in an adjacent garden. Fletcher suggested the furnace had a short working life, since the masonry was little worn and the slag heap was comparatively small. Marshall and Davies-Shiel (1969, 234) found slag at NY 020137, on the north side of the entrance lane near Ehen Hall corn-mill. They pointed out that pig iron was exported from west Cumberland by Richard Patrickson of Cleator in 1700–01 and that he had been experimenting with a mixture of coal and charcoal as fuel. Patrickson died in 1705, which may have brought the furnace to an end (Phillips 1977, 3, 27). More recently, Peter Brown has investigated both the site and its history, apparently establishing that Cleator was the first furnace in Great Britain to use mineral fuel for smelting, if only experimentally (Cherry 1982, 226; Blick 1984, 49).

Cunsey, Lancs.   SD 383937 [96]

A furnace was erected on the Cunsey Beck west of Lake Windermere in 1711–12, on or near the site of a bobbin-mill that later became a joinery works. Part of a nearby house retains the thick wall of the furnace stack and there is a well constructed weir and head-race by Eel House Bridge to the west. Cunsey furnace was built by the Cheshire ironmasters Edward Hall and Daniel Cotton, both of whom appear in the Knights' accounts as suppliers of pig to the Stour Partnership from Cunsey between 1746 and 1755 (Ince 1991, 117). The furnace, which probably went out of use about 1755, appears in the list of closures between 1750 and 1788 with Kendall as occupier and the marginal comment 'down' (Fell 1908, 209; Schubert 1957, 373; Barnes 1951, 84; Marshall and Davies-Shiel 1969, 37, 237–8).

Duddon, Cumberland   SD 197883 [96]

In 1736 the Cunsey Co. established a furnace at Duddon Bridge, which ultimately came into the hands of Harrison Ainslie & Co. (Fell 1908, 215–16, Barnes 1951, 84; Schubert 1957, 374; Morton 1962; the latter lists earlier references). The furnace passed from the original Cunsey Co. partners to Hall, Kendall & Co., Kendall, Latham & Co. and finally Joseph and Richard Latham, who sold the works to Harrison Ainslie in 1828. Duddon appears in the Knights' Stour Partnership accounts as a supplier of pig in 1740–2 and 1749–57 (with Hall named as the tenant) and 1773–4, 1783–6 and 1792–9 (then in the hands of William Latham) (Ince 1991, 117).

In 1796 Duddon was said by the Excise to make 32 tons of pig a week; the trade's own estimate was 400 tons a year, which may be a figure based on actual production or may be intended to represent eight tons a week, closer to the 10 or 11 tons which the furnace was actually producing in the mid-eighteenth century. In 1806 the output was returned as 175 tons, which may be a rather more realistic figure. Like its neighbours at Backbarrow and Newland, as well as Lorn (i.e. Bonawe) in Argyllshire, Duddon survived into the era of *Mineral Statistics*, where the furnace is listed as being in blast in 1854–5, 1857 and (for the last time) 1871. Truran (1855) claims an improbable 1,000 tons p.a. as the output at Duddon.

After lying derelict and slowly deteriorating for many years, the furnace has finally been accorded the official recognition (and protection as a scheduled monument) which it deserves as the outstanding site of its kind in England and, with Dyfi in Cardiganshire and Bonawe, one of the three most impressive in Britain. During the 1980s the Lake District National Park Authority undertook a major programme of excavation and consolidation at Duddon (Cherry 1982, 226–8; Egan 1985, 185), the fruits of which can now be seen by visitors. The site is readily accesible from the minor road which leaves the A595 immediately west of Duddon Bridge to run north alongside the River Duddon. The furnace itself is substantially complete, with the casting-house on the south side, the blowing-house to the east and the wheel-pit to the north. To the west, it is possible to walk up to the charging level of the furnace on a modern footway, which runs above store-rooms and offices below.

The furnace was powered, at first by bellows, later by cast-iron cylinders, using water brought in a leat running south from the River Duddon, which ends at a right-angle to the wheel-pit. Just above the wheel-pit the water supply was supplemented by a stream coming down the hillside from the west. A little way up the hill to the west of the furnace itself stand two very substantial charcoal stores and a rather smaller iron ore store. South of the footpath which provides access to the site (and outside the area currently leased by the National Park) there is a derelict manager's house and outbuildings.

Above: **Backbarrow.** The much rebuilt furnace that was the last in Britain to use charcoal until its conversion to coke in 1921. *Below:* **Duddon,** looking east across the site, with an iron-ore store in the foreground and the furnace beyond.

Frizington, Cumberland   NY 030178 [89]

Although not conducted using a blast furnace, William Wood's experiment with coke-smelting at Frizington in 1728 may conveniently be included here. The site of the reverberatory furnace used by Wood was near Bleak House on the lane to Arlecdon, as shown by a sketch-plan in the Lowther MSS in the Cumbria Record Office at Carlisle. The experiments were costly and not very successful (Marshall and Davies-Shiel 1969, 37–9, 243; cf. Flinn 1961–2 and Treadwell 1974; see also Bellingham (Northumberland) for Wood's activities elsewhere in the north of England). Fletcher (1881, 14–15) reported the survival of two circular blast furnace bases at Howth Gill, Frizington, which he concluded were remains of a later ironworks than John Wood's, and suggested that Wood's experiments were conducted either on the same site or lower down the same gill, or on a nearby stream called Dub Beck.

Halton, Lancs.   SD 5064 [97]

The 1794 list includes a charcoal furnace at Halton on the River Lune, two miles NE of Lancaster, said to have been built in 1756, owned by a Mr Bradshaw and occupied by 'Hatton & Co.', which is possibly an error for Halton Co. The plant consisted of a single water-powered furnace. The date 1754 appears in the margin of the same line of the list near the spaces ruled for mill plant, but there was apparently only a furnace there.

Further light is shed on the origins of this works by a note in Halton parish register that 'In the year of our Lord 1752 Halton Furnace was erected, and they used [in] the building thereof nigh fifteen thousand cart load of stones; at twopence-halfpenny per cart' (Pape 1959). The following year an agreement was made between Miles Postlethwaite for the proprietors of Leighton furnace (q.v.) and Miles Birkett for Halton as to the purchase of local woods which were then for sale (Fell 1908, 142), indicating that the latter furnace was in use by then. An ironmaster named John Ayrey was supplying the Knights' Stour Partnership forges with pig from Halton between 1753 and 1781, succeeded by Samuel Routh in 1781–3 (Ince 1991, 117).

In 1808 a notice in the *Lancaster Gazette* announced that the sixteen proprietors of the Halton Iron Co. had dissolved their partnership and stopped work at the 'foundry'. This perhaps implies the end of smelting some years before, since Halton does not appear in the lists of furnaces of

1796 and 1806. The advertisement of 1808 named receivers who were to collect and pay all the debts of the company and to manage the business until existing stocks had been 'wrought up and disposed of'. These included 'charcoal, bar iron of every description, cast iron work to any pattern ... bills, hoes, spades and edge-tools of all sorts'. Eventually Thomas Butler Cole of Beaumont Cote purchased the ruined furnace and built a mill on the site. After that failed, he pulled down most of the buildings, enclosed the ground and set up a gateway with the intention of making a private road to the Cote (Pape 1959). There are thus no surface remains of the ironworks visible today.

Holme Chapel, Lancs.　　　　　　　　　　　　　SD 879280 [103]

About 1700 the first blast furnace in Lancashire was built at Holme Chapel, between Todmorden and Burnley. It is first referred to as 'Mr Wilmott's furnace' and from 1702–3 onwards small quantities of pig were received from it by Colnbridge and Kirkstall forges. Robert Wilmott married Mary, one of the five daughters of the elder Thomas Dickin; these five eventually became coheiresses to the partnerships of their brother when he died without issue in 1701. Robert Wilmott was probably the son of James Wilmott of Hartlebury (Worcs.), who was associated with the iron trade of the lower Stour valley: he set up Wilden forge and converted the family's fulling-mills at Mitton into a forge. Holme Chapel furnace had an unfortunate history: in 1713 three ironmasters from the Barnsley area joined Wilmott in running it, John Silvester of Burthwaite Hall, Nicholas Burley of Woolley and John Spencer of Cannon Hall, who intended to build a forge and slitting-mill. In the event the furnace operated unprofitably in the first half of the eighteenth century, seeking to supply the Lancashire market from a nearer base than Colnbridge in the West Riding (Awty 1957, 90).

Schubert (1957, 378) made the improbable suggestion that Holme Chapel had late sixteenth-century origins and then relies on Raistrick (1938) for early eighteenth-century references to the Spencers' activities there, suggesting the furnace closed about 1750. It does not appear in the list of closures between 1750 and 1788.

Recent field investigation by Lancaster University (Ponsford 1991, 164), together with work by Mr R. Redfern and Mr T. Thornber (to whom I indebted for this information), has located slag in the bed of the adjacent River Calder, part of a leat and a wheel-pit at the site, which later in the eighteenth century was used as a pottery before landscaping in the ninteeenth century created the present Cliviger Ponds there.

Leighton, Lancs. SD 485778 [97]

In 1713 the Backbarrow Co. built a furnace at Leighton near Arnside, which in 1755 was taken over by the Halton Co. of Lancaster and remained in use until 1806 (Fell 1908, 209–11, Barnes 1951, 84; Schubert 1957, 379). This is confirmed by the 1794 list, where Leighton appears (wrongly) under Cumberland & Westmorland, with the owner given as Lord Derby and the occupiers 'Hatton & Co.' (possible in error for Halton Co.; cf. Halton). Located two miles from Milnthorpe, the works consisted of a water-powered charcoal furnace built in 1715. In 1796 output at Leighton was said to be 780 tons a year, probably meaning in fact 15 tons a week; the furnace was said to be in blast in the survey of 1806, but no output was given. Peat was used as a fuel at Leighton in the first half of the eighteenth century (Morton 1965). James Machell appears in the Knights' Stour Partnership accounts as a supplier of pig from Leighton in 1748–58 and 1763–4 (Ince 1991, 117).

The site of the furnace (of which there are no surface remains) is indicated today by a plaque on the roadside near the entrance to a private residence created from the large charcoal store mentioned by Marshall and Davies-Shiel (1969, 39, 45, 249).

Little Clifton, Cumberland NY 059279 [89]

A furnace was set up on the River Marron in 1723 by the Cookson family of Durham (Marshall and Davies-Shiel 1969, 37, 39, 41, 250–1; Lancaster and Wattleworth 1977, 19–20; cf. Chester-le-Street, Durham). This was an early and unsuccessful attempt at coke-fired smelting, which has left few traces on the ground. The furnace was still in use in 1765, when it was visited by Gabriel Jars, who reported that it could not make good forge pig even with charcoal. Fletcher (1881, 9–10) suggested that the furnace was abandoned in 1781, if not before, when Cookson's colliery in the same area was drowned out. He concluded that 'with an imperfect knowledge of fluxes and the feeble pressure of blast in use at that time, it was not practicable to smelt in a satisfactory manner the red haematite ore of West Cumberland in coke furnaces'.

Above: **Leighton:** The former charcoal store now converted into a private residence.
Below: **Nibthwaite.** The base of the furnace, showing the casting-arch.

The site was about three-quarters of a mile south of the village of Little Clifton, near the bridge carrying the lane from Little Clifton to Dean over the river. About a quarter of a mile to the west, on another lane, a farmstead named Furnace House still stands.

Low Wood, Lancs. SD 346836 [96]

A furnace and forge were erected on the River Leven near Backbarrow in 1748 by a partnership which included Isaac Wilkinson, Job Rawlinson, George Drinkal and William Crossfield (Fell 1908, 218–20, Barnes 1951, 85; Schubert 1957, 380; Marshall and Davies-Shiel 1969, 45). Low Wood probably appears in the list of closures between 1750 and 1788 as 'Larswood', with the occupier given as 'Sunderland' and the marginal comment 'stands'.

This corresponds with the details given by Fell, who states that by 1760 only one of the original partners (Crossfield) was still connected with the works and that the others had been succeeded by John Sunderland of London, Thomas Sunderland of Bigland and John Wilson of Hawkshead. Thomas Sunderland died in 1774 and, after the death of John Sunderland in 1782, the lease was assigned to the Backbarrow and Newland companies. They used the furnace for a few years and then abandoned the works in 1785; the site was subsequently occupied by a gunpowder-mill and there are no visible remains of the furnace. The Knights' Stour Partnership accounts confirm the chronology: John Sunderland supplied pig from Low Wood in 1761–78, followed by Thomas Sunderland and John Machell in 1784–7 (Ince 1991, 117).

Maryport, Cumberland NY 034362 [89]

The furnace at Maryport was built under a fifty-year lease of 1752 by Humphrey Senhouse to James Postlethwaite of Cartmel, William Lewthwaite of Kirkby Hall, William Postlethwaite of Kirkby and Thomas Hartley, John Gale, Edward Tubman and Edward Gibson, all of Whitehaven, who were given licence to build a forge, furnace and associated works. Exactly when the furnace was built is not clear; the date '1763' on a manager's house may point to a delay in work beginning or simply a date at which profits permitted this additional building. The furnace was at work making castings in 1777 and in the same year iron blowing cylinders, made at Bersham, were installed, presumably replacing leather bellows.

The enterprise began to falter shortly after this date, probably in part for lack of sufficient water-power and lack of capital with which to install a steam-engine that would have overcome the first problem. In May 1783 the works were advertised for sale by auction; they appear to have been withdrawn but in January the following year the site was recovered by the ground landlord. It was not re-let and the furnace stood idle, surviving until its very unfortunate demolition in 1963.

It is not clear whether Maryport was built explicitly as a coke furnace. The sale notice of 1783 states that it could be blown with either charcoal or coke fuel; in 1765, when Jars visited the area, he reported that one furnace near Workington was then in blast and a second one was in course of construction, the latter intended to be a coke furnace. This must have been the Seaton furnace (q.v.), so that Maryport was evidently then using charcoal. On the other hand, a bank of seventeen beehive coke-ovens, which survived the demolition of 1963, were built on the site, apparently when it was first established, which points to an attempt to improve the method of coke-making to enable mineral fuel to be used. Schubert (1952; 1957, 381) twice drew attention to the remarkable survival of this important site, but to no avail; Tylecote *et al.* (1963) made a detailed examination of the remains after demolition and assembled documentary evidence for its history; Marshall and Davies-Shiel (1969, 40–1, 253) describe what could be seen there six years afterwards.

Maryport appears as 'Netherall' (i.e. Netherhall) in the list of closures between 1750 and 1788, with the occupier's name given as Atkinson (as at Seaton) and the marginal comment 'down'. This corresponds with the evidence of the Senhouse papers, which name the last occupiers as J. Atkinson & Partners, and with the Knights' Stour Partnership accounts, where Hartley, Atkinson & Co. appear as suppliers of pig from Netherhall in 1777–80 (Ince 1991, 117).

Netherhall: see Maryport

LANCASHIRE AND CUMBERLAND

Newland. *Above:* The furnace (scaffolded for repairs) and blowing-house, with marks indicating the position of a roof over the wheel-pit. *Below:* The charcoal store.

Newland, Lancs.   SD 299798 [96]

In 1746 Richard Ford, together with his son William, Michael Knott and James Backhouse, built a furnace at Newland, about a mile NE of Ulverston. It was rebuilt in 1770; a forge was added in 1783 and a rolling-mill (the first in the district) in 1799. Production ceased in about 1891 and the works were partially dismantled in 1903 (Fell 1908, 217–18; Barnes 1951, 85; Schubert 1957, 382; Marshall and Davies-Shiel, 42, 44, 263).

In the 1794 list the furnace appears as 'Newlands', 17 *(sic)* miles from Ulverston, owned and occupied by the executors of G. Knott, with the plant consisting of a water-powered furnace said to have been built in 1750, a finery and a chafery. The Stour Partnership accounts confirm the ownership: George Knott supplied pig to the Knights from Newland in 1781–5, his executors from then until 1796 (Ince 1991, 117). In the list of ironworks of 1796 an output of 700 tons was returned for Newland; ten years later the furnace was listed again but with no output shown. The works survived to be listed annually from 1854 in *Mineral Statistics*, with Harrison Ainslie & Co. as owners. The furnace was generally in blast every year down to 1890, which, according to the official returns, was the last year of operation. Truran (1855) prints an output figure of 1,200 tons p.a. for Newland, which seems an improbable guess.

Today, there is still a good deal to be seen at the site, which forms a self-contained hamlet on the south bank of the Newland Beck, just off the A590 close to the bridge carrying the main road over the stream. The property is privately owned: prospective visitors should first call at the house on the right-hand side of the lane beyond the furnace, which was once the manager's home. The furnace itself is substantially intact and is currently being cleared and consolidated under the auspices of the Cumbria Industrial History Society. The remains of both the casting-house and blowing-house can be seen alongside, together with the wheel-pit. On the higher ground beyond the furnace stands a large charcoal store.

Nibthwaite, Lancs.   SD 295883 [96]

The furnace at Nibthwaite on the River Crake was built in 1735, initially financed by Richard Ford and Thomas Rigg but really the foundation of the Newland Co. which later, as Harrison Ainslie & Co., was to acquire all the charcoal ironworks in Furness. According to Fell the furnace closed as early as 1755, but the forge survived until 1840. (Fell 1908, 211–15; Barnes 1951,

84; Schubert 1957, 382; Morton 1963; Marshall and Davies-Shiel 1969, 40, 256.) Ford supplied pig to the Knights' Stour Partnership in 1740–42 (Ince 1991, 118).

In the list of furnaces closed between 1750 and 1788 Nibthwaite appears with Ford as the occupier and the marginal comment 'forge'; this corresponds with the entry in the 1794 list, in which two fineries and a chafery (but no furnace) are said to be owned and occupied by G. Knott. After the forge closed the works became a bobbin-mill and later a carpenter's and undertaker's premises.

During the 1980s, when the main workshop was converted into a private residence, excavations by David Cranstone (Egan 1985, 185) revealed that much of the furnace remained intact beneath later buildings; the structure is now a scheduled monument. The results of Cranstone's work can be seen, on application to the owners of the property, incorporated into the garage beneath the house. The head-race from the Crake is also well preserved, whilst the former charcoal store, which stands on the right on the private drive which gives access to the whole group of properties, has likewise been converted for residential use.

Penny Bridge, Lancs. SD 308838 [96]

In 1748 local wood-owners, after much disagreement with the ironmasters, decided to erect their own furnace at Penny Bridge under the leadership of William Penny of Penny Bridge. The company later amalgamated with the Backbarrow Co. and the Penny Bridge furnace was last used in 1780. It was demolished in 1791 and a flax-mill built on the site in 1805, leaving no visible trace of the ironworks (Fell 1908, 221–2; Barnes 1951, 86; Schubert 1957, 383; Marshall and Davies-Shiel 1969, 40, 243). In the list of furnaces closed between 1750 and 1788 Penny Bridge appears with 'Machel' as occupier and the marginal comment 'stands'.

Seaton, Cumberland NY 013295 [89]

A furnace was established at Barepot, near Seaton, about a mile NE of Workington on the north bank of the River Derwent, about 1762, and was visited by Jars in 1765. He apparently saw the charcoal furnace at Maryport (q.v.) and reported that a coke furnace was also being built in the locality, presumably meaning that at Seaton. In 1779, according to a letter quoted by

Fell, there were two forges and a furnace on the site. Seaton, however, appears in the list of closures between 1750 and 1788, as a charcoal furnace, immediately after 'Netherall' (i.e. Maryport), likewise occupied by Atkinson, and said then to 'stand'. The Knights' Stour Partnership accounts supply the further detail that in 1771–7 Samuel and Sampson Freeth supplied charcoal-smelted pig from Seaton, as did Richard Dearman & Co. in 1779–89 and J. Petty Dearman in 1793–5 (Ince 1991, 118).

The works appears as a water-powered coke-fired furnace in the 1794 list, occupied by the Seaton Co., with a forge consisting of three fineries, a chafery and a slitting-mill. The ground landlord was named as 'Mr Christian' and the date of construction for the furnace was given as 1760. In 1796 the Excise suggested an output of 40 tons a week for Seaton, which the trade reduced to 24 tons; in 1806 an output of 670 tons p.a. was returned. Both in the latter year and in 1810 the tenant was named as Spedding & Co. The compiler of the 1825 list was unable to obtain a return for Seaton, which he believed to be out of use. According to Fletcher (1881, 12–13) the works were established by Hicks, Spedding & Co. under a 99-year lease granted by Sir James Lowther. Fletcher believed that the furnace was last used in 1857, having been acquired by Tulk Ley & Co. in 1837, whereas a modern account suggests that only a foundry, the rolling- and slitting-mill and engine-works continued to operate until the mid-nineteenth century, when the site became a tinplate works (Marshall and Davies-Shiel 1969, 39, 42–3, 45–7, 259).

There are some nondescript nineteenth-century industrial buildings at the site today but no visible remains of the furnace itself.

Wilson House, Lancs. SD 426809 [96]

Wilson House Farm, at Lindale-in-Cartmel, is supposed to be the site of coke-smelting experiments by Isaac and John Wilkinson *c.* 1749. In 1969 Marshall and Davies-Shiel (p. 250) located slag there and an abandoned wharf on the River Winster, although there is little of this to be seen today. At the farmhouse, however, there are numerous eighteenth-century cast-iron pipes, apparently for use at the Chelsea and Paris waterworks, and some used to support a barn roof with 'WILKINSON 1784' cast into them (ex inf. A. Lowe).

- Bellingham

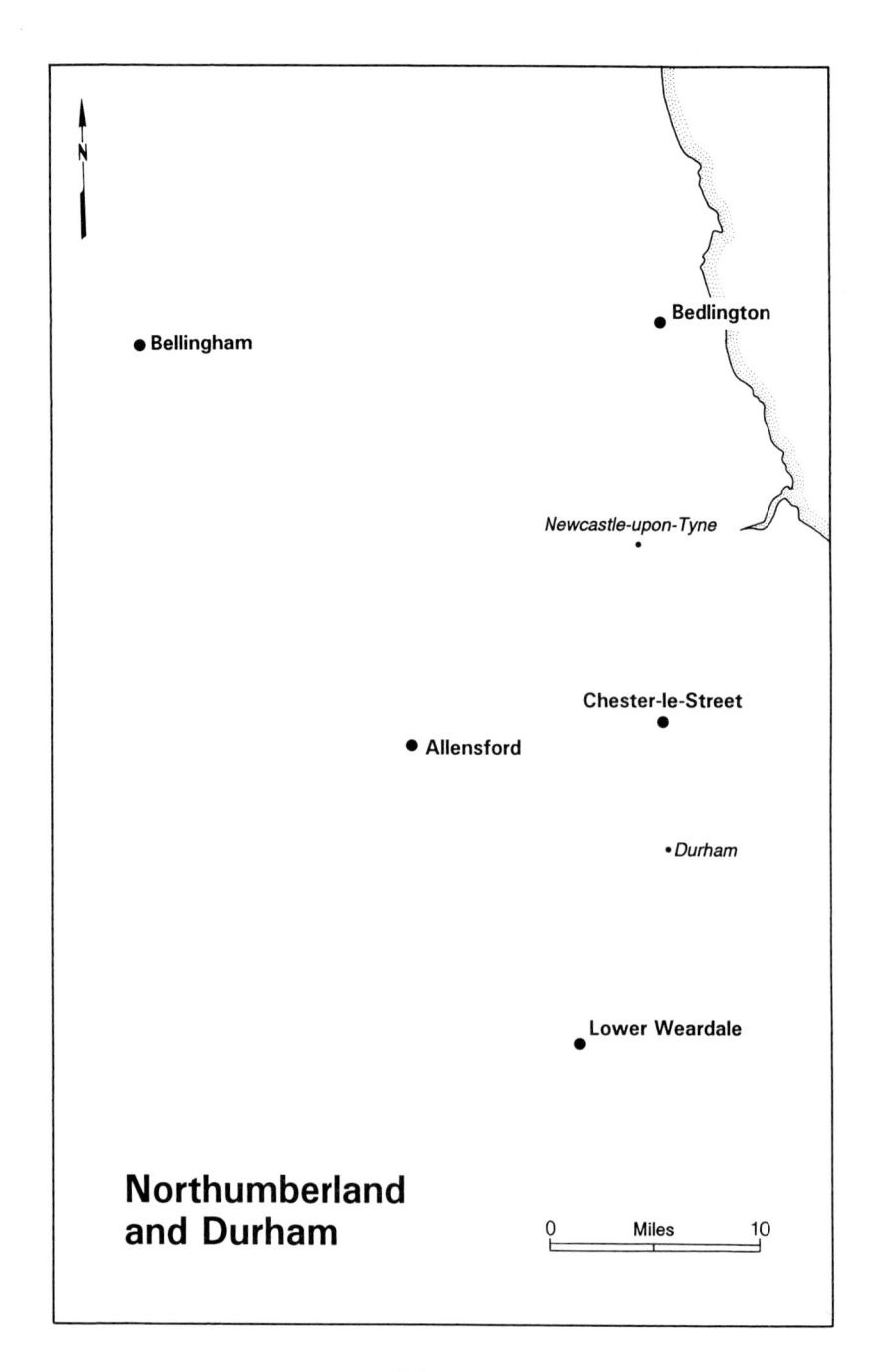

• Bedlington

Newcastle-upon-Tyne
•

Chester-le-Street
•
• Allensford

• Durham

• Lower Weardale

**Northumberland
and Durham**

0   Miles   10

10

Northumberland and Durham

Compared with most other regions, the early history of blast-furnace ironsmelting in the North East is poorly documented. Much of what is known at present derives from the work of the ironmaster-antiquary Isaac Lowthian Bell in the 1860s, apart from a survey by R.F. Tylecote (1983). Schubert (1957) failed to include any sites in Northumberland and only one in Durham, of which he provided a sketchy account.

In addition to the sites described below, the first edition of this work included an entry for Derwentcote, Co. Durham, where Atkinson (1974, 187–8) suggested that there might have been a blast furnace in addition to the forge and steelworks. There is no early historical evidence for this, nor were any remains of a furnace discovered during the extensive campaign of excavation and consolidation carried out during the 1980s at what is now an important guardianship monument (Egan 1989, 61–4).

Allensford, Northumberland   NZ 076505 [88]

The *History of Northumberland* (VI (1902), 301) cites the muniments of Sir Arthur Middleton of Belsay Castle as the source for references to ironworks at Allensford, near Consett, on the north bank of the Derwent. A deed of 1673 refers to an 'Iron-forge, for the making and working of iron', as do others of 1683 and 1687, while in 1692 a blast furnace and forge were leased to the Yorkshire ironmaster Dennis Heyford, having previously been held by a man named Davison. In 1713 a forge (only) was conveyed to Nicholas Fenwick of Newcastle. The furnace does not appear either in the 1717 list or any later survey; the forge is included in the lists of 1736 and 1750 (Hulme 1928–9). The furnace is also referred to in a volume of poems by Joshua Lax, published in 1884, when it was suggested that the site had been operated by the Bertrams, one of a group of German families who settled around Shotley Bridge in 1688–9. It is presumably the furnace described by Bell (1864, 83) as a 'small high-blast furnace', five or six feet high in the boshes, remains of which were then still visible.

The site was investigated by Linsley and Hetherington (1978), who excavated remains of the furnace and a calcining kiln. Their historical notes are drawn mainly from the County History; the deeds of 1673 and 1713 are now Northumberland RO, ZMI.B.2/V/8 and ZMI.B.1/XII/1 respectively, those of 1683 ZMI.B.2/V/12 and ZMI.B/VI/2. They add that the forge was leased to the Crowleys in 1728, presumably as successors to Fenwick. At present, all that can be said about smelting here is that a blast furnace is mentioned in 1692 but not in 1673 or 1713, when there appears only to have been a forge. The site itself was also described by Tylecote (1983, 93–4).

Bedlington, Co. Durham, later Northumberland NZ 276820 [81]

The early history of the ironworks on either side of the River Blyth at Bedlington has recently been clarified by Chris Evans (1992), whose account has swept away the confusion of earlier writers (*History of Northumberland*, IX (1909), 298ff.; Bergen *c*. 1948, 4–9; Martin 1974, 3–7; Tylecote 1983, 99).

Although a forge was established on the south (Bebside) bank of the river by a Newcastle merchant named William Thomlinson under a lease of 1736, there was no attempt at smelting in the area until after William Maling of Sunderland (1698–1765) acquired a site on the opposite bank (in Bedlington township itself, then a detached outlier of Co. Durham) in 1759 (rather than Martin's date of 1757: ex inf. S.M. Linsley). The earliest evidence that a furnace was in use comes from an advertisement in *Aris's Birmingham Gazette* (3 Nov. 1766), which offered shares in Bedlington works and stated that the furnace is 'for smelting Iron from the Mine by Pitcoal or Charcoal' and is 'Well-established in a Vend for Cast Ware, &c.' (Ashton 1924, 37–8). A few years later, a local topographer (Wallis 1769, I.125) described Bedlington as 'The only iron-work of any eminence with us at present', where 'The ore is digged out of the hanging banks by the river with great labour and pain', presumably then to be smelted nearby. In 1778 the *Newcastle Courant* referred to a blast furnace, air furnace, casting-house and other plant at Bedlington, suitable for smelting and manufacturing iron from the ore.

Smelting evidently ceased at Bedlington sometime between 1778 and 1788, when a version of the list of charcoal furnaces closed since 1750 preserved among James Weale's papers in the Science Museum Library lists Bedlington as 'entirely declined'. It does not appear in the Boulton & Watt text of the same list, although it was included there among the unsuccessful coke-fired furnaces appended to the main list of charcoal furnaces, which further establishes that both fuels had been used at Bedlington. In 1788,

Hawks, Longridge & Co., who already had the works on the south bank of the Blyth, took over the Bedlington site as well. This is confirmed by the 1794 survey, in which two works are listed at Bedlington (i.e. those on either side of the river), both occupied by Hawks & Longridge, one owned by the Bishop of Durham and the other by a 'Mr Ward', the plant consisting of a rolling-mill on one bank and a slitting-mill on the other. Evans has established that the latter was at Bebside and the former at Bedlington; the furnace was still standing in 1789, when it was assessed to land tax, but later assessments list only a foundry. Longridge & Co. continued to occupy the site into the nineteenth century, when Bedlington became a well-known manufacturer of railway goods.

Bellingham, Northumberland NY 861796 [87]

Wallis (1769, I, 125), in discussing the small Northumberland iron industry, recalled that:

> There was some years ago an iron-work at Lee-Hall, on the edge of the river of North Tyne, near Bellingham ... It was under the direction and conduct of Mr Wood, son of Mr Wood, famous for being the projector of the halfpence and farthings for Ireland by patent. He made a good deal of bar-iron, but charcoal becoming scarce, he removed to Lancashire, where he attempted to make it with pit-coal.

Bell (1864, 63) claimed that, although Wallis mentioned only bar iron, 'there is no doubt from the remains still existing, that Wood also produced pig iron there'. No remains of any kind are mentioned by Atkinson (1974) at Bellingham, nor is an ironworks referred to in the *History of Northumberland* article on the township (XV (1940), 222 *et seq.*). Neither of the two most recent accounts of the ironmaking activities of William Wood (Flinn 1961–2; Treadwell 1974) mentions any connection with the North East, although much of his career, especially before he attempted to establish coke-smelting at Frizington (in Cumberland, not Lancashire, q.v.) in 1728 remains obscure.

The Bellingham site was investigated by Tylecote (1983, 90), who found the possible remains of a blast furnace, together with a quantity of what he concluded was bloomery slag. Later blast furnace slag had probably been brought on to the site. At present, however, there appears to be insufficient evidence to establish exactly what the 'ironworks' mentioned by Wallis consisted of.

Chester-le-Street, Co. Durham　　　　　　　　　　NZ 265518 [88]

In 1864 Isaac Lowthian Bell told the British Association that 'Mr Joseph Cookson, in a very interesting document drawn up for the writer, mentions that, for the Whitehill furnace, built in 1745, and abandoned before the end of the last century, ironstone was gathered in Robin Hood's Bay, and conveyed by water to Picktree, on the Wear, near Chester-le-Street, and carted from that place to the works'. A little later in the paper he goes on to say that:

> To Mr I. Cookson, who had recently purchased the Whitehill estate, near Chester le Street, the merit belongs of erecting and working the first blast furnace with coked coal in the North of England. The Whitehill furnace was 35 feet high, 12 feet across the boshes, and produced 25 tons of iron per week. The blast was supplied by a bellows, worked by a water wheel placed on Chester Burn. Its mode of supply of ironstone was from the thin bands on Waldridge Fell, and from Robin Hood's Bay, as has been already mentioned. The coal, of course, was obtained from the immediate locality. Mr Joseph Cookson, a descendant of the founder of pit coal smelting in this district, has given many curious particulars of this early attempt. The iron was used for colliery castings, and latterly for Government ordnance. Frequent interruptions, for want of water to drive their wheel, led at length to the furnace being 'gobbed', and ultimately abandoned, about the close of the last century (Bell 1864, 79, 84).

These passages have been quoted at length since they appear to be the earliest authority for one of two pre-1750 attempts at coke-smelting in the North East.

The story reappeared in *VCH Durham* (II (1907), 290), with the additional comment that 'The foundry was begun early in the eighteenth century', for which the authority was a footnote saying that 'Mr N.C. Cookson thinks that the blast furnaces at Whitehill were started about 1704', which in turn is the basis of later suggestions that the works dated from as early as 1704 and had more than one blast furnace; both ideas seem highly improbable. *VCH* also printed, from an original in the hands of the Newcastle antiquary Richard Welford (died c. 1913), part of a deed of 1760, of which a fuller abstract was printed elsewhere about the same time (*Proc. Soc. Antiquaries of Newcastle-upon-Tyne*, 3rd series, 3 (1907–8), 170–1; cf. Durham University, Dept of Palaeography & Diplomatic, Cookson MSS, Box 3/48). The deed, the original of which has not been located, recites earlier transactions going back to 1721 and the establishment of the blast furnace at Little Clifton, Cumberland (q.v.), in 1723 (for the original deed of 1721 see Gateshead Public Library, Cotesworth MSS, CA/2/65). A partnership deed of 1729 is also recited, which refers to foundries at Clifton, Gateshead and Newcastle, as well as an agreement for a lease of land at Whitehill, Co. Durham, on which a blast furnace had been built (and was evidently still in use in 1760). Unfortunately, the date of this agreement is not given in the published abstract of the deed of 1760. *VCH* repeats Bell's comments about the sources of coal and ironstone for the furnace, with the added embroidery that 'at one time the Government got most of their ordnance from the forge at Whitehill'. The Cooksons are also said to have made a fortune from ordnance contracts during the Napoleonic War and to have supplied a cast-iron safe to the churchwardens of Chester-le-Street in 1813.

At present there appear to be only two pieces of firm evidence for the continued existence of the Whitehill furnace after 1760. One is a letter from John Cookson to Isaac Wilkinson of June 1766 (Tyne & Wear Archives, 1512/5571; I am very grateful to Chris Evans and David Cranstone for drawing this important new evidence to my attention), in which Cookson returns answers to a number of queries from Wilkinson concerning the Whitehill furnace and, in a postscript, refers to cast-iron blowing cylinders, which Cookson was apparently thinking of installing at Whitehill and which were to be powered by a water-wheel 27ft in diameter. Cookson invited Wilkinson to come and stay at Whitehill, presumably while Cookson considered adopting Wilkinson's blowing cylinders, patented in 1757 (Chaloner 1960, 39–40). It is not clear whether cylinders were eventually installed in place of bellows at Whitehill.

Secondly, towards the end of its life, the furnace at Whitehill did make small quantities of cannon for the Board of Ordnance, although Cookson was neither a major supplier, nor a particularly successful one. He was only a contractor from 1777 and it is clear that the board used him only in times of great need; in 1780–1 he was responsible for just under 10 per cent of the cast-iron cannon bought by the Ordnance (Brown 1988, 106–8; cf. Tomlinson 1976, 388). When precisely Whitehill was abandoned is not certain but it seems definitely to have been by 1788, since both Chester-le-Street and Bedlington (q.v.) appear in the list of failed coke-fired furnaces of that year.

The history of the site is further clarified by Hutchinson (1787, III, 398n.), who refers to a coke blast furnace then at work at Whitehill near Chester, whose dimensions (about 34ft high and 12–13ft wide at the widest part) and operations are described. The same note adds: 'About three miles west of Chester is a place called the Old Furnace, where very lately was to be seen the bottom of a furnace hearth, according to the usual method of building them now, but of much smaller dimensions: They had blown the bellows with a water wheel, as appears by the cut of a water race to convey it to the wheel from an upper part of the burne'. This appears to be a reference to a disused blast furnace some distance from Whitehill (possibly on the Twizell Burn around NZ 2251). Tylecote (1983, 91–2) located the later site, where 'Furnace Farm' now stands near the centre of Chester (NZ 265518), but was inclined to reject the idea that there were two separate furnaces in the neighbourhood. On the basis of Hutchinson's comments, however, it seems possible that the 'Old Furnace' preceded that at Whitehill, which Hutchinson apparently saw in use, even though it fails to appear in the otherwise generally accurate 1794 list.

Finally, Dr Linsley (pers. comm.) suggests that the Whitehill furnace may date from 1747 rather than 1745, since it was in the former year that the Cookson family acquired the estate on which it stood.

Lower Weardale, Co. Durham *c.* NZ 2029 [93]?

A stray account for 1664 establishes that John Hodgshon was then operating a blast furnace for the Bishop of Durham somewhere in the county. The furnace is not located but there are three place-names in the account, Hunwick, Bedburn and Birtley (ex inf. R.C. Norris, Durham University Library). There was ironstone stored at Hunwick Moor in 1664, which suggests that the furnace was probably in lower Weardale, possibly somewhere near Bishop Auckland. The furnace apparently has no other history; there seems no warrant for *VCH Durham* (II (1907), 280–1, where the account was printed from Durham Cathedral Library, Mickleton MS 91, item 29, whence Schubert 1957, 374) to link the document with references to spoliation of woodland in Weardale in 1629 or ironstone mining at Chester-le-Street three years before that.

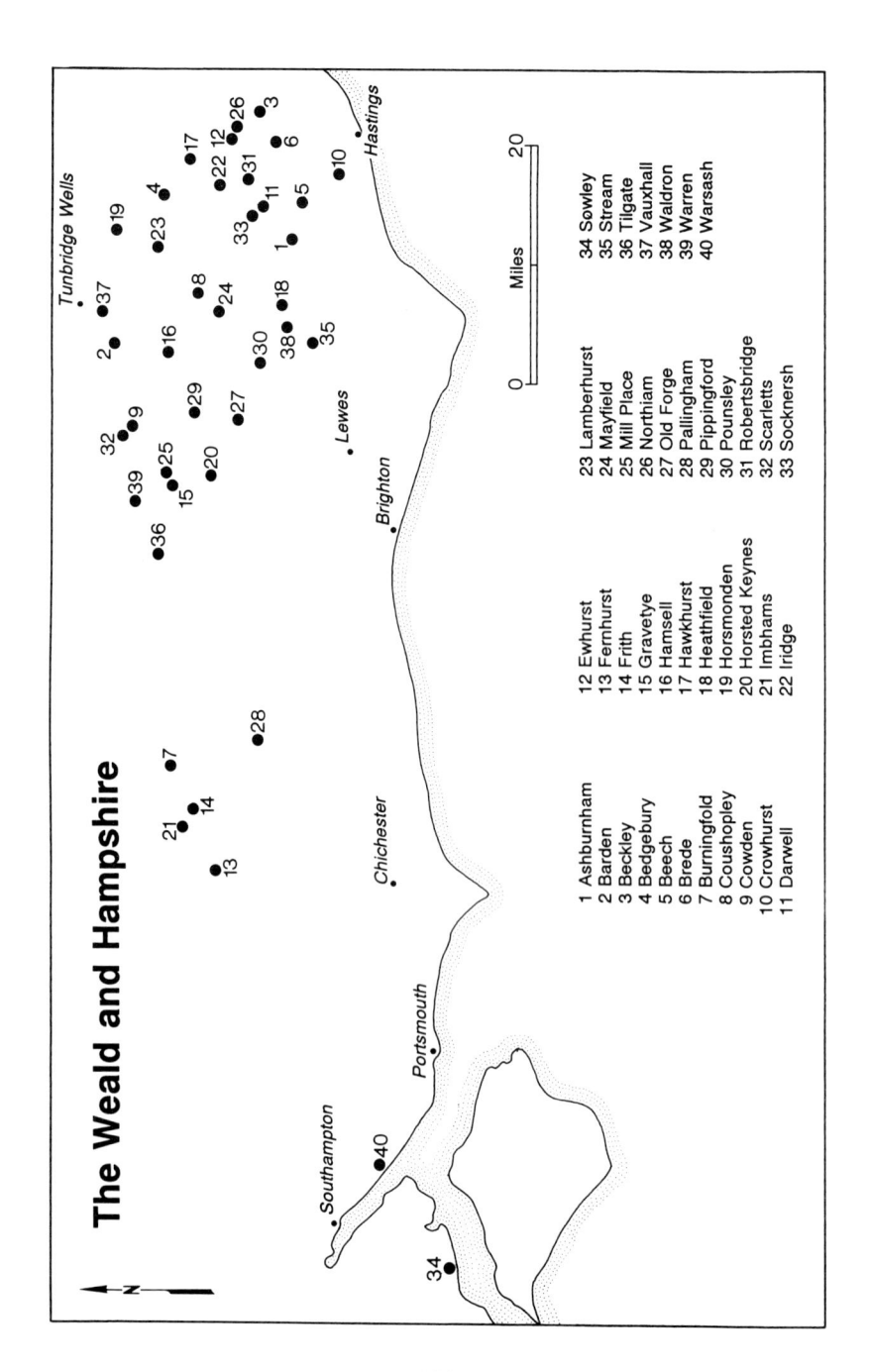

The Weald and Hampshire

1 Ashburnham	12 Ewhurst	23 Lamberhurst	34 Sowley
2 Barden	13 Fernhurst	24 Mayfield	35 Stream
3 Beckley	14 Frith	25 Mill Place	36 Tilgate
4 Bedgebury	15 Gravetye	26 Northiam	37 Vauxhall
5 Beech	16 Hamsell	27 Old Forge	38 Waldron
6 Brede	17 Hawkhurst	28 Pallingham	39 Warren
7 Burningfold	18 Heathfield	29 Pippingford	40 Warsash
8 Coushopley	19 Horsmonden	30 Pounsley	
9 Cowden	20 Horsted Keynes	31 Robertsbridge	
10 Crowhurst	21 Imbhams	32 Scarletts	
11 Darwell	22 Iridge	33 Socknersh	

130

11

The Weald and Hampshire

The Wealden iron industry, the subject of Straker's famous monograph of 1931, was re-examined a few years ago in a volume that will probably remain a standard work for as long as its predecessor (Cleere and Crossley 1985). The entries here for furnaces which appear to have operated after 1660 are based directly on this work, with cross-references back to Straker and Schubert. The availability of Cleere and Crossley's work has obviously lightened the load of preparing this gazetteer considerably; in particular the National Grid references (which Straker did not use) are an enormous boon to all future work in this field. I have also included two charcoal furnaces in south Hampshire (Sowley and Warsash) in this chapter; they do not, of course, form part of the Wealden industry strictly speaking.

I am very grateful to Jeremy Hodgkinson for considerable help in revising this chapter for this edition.

Ashburnham, Sussex TQ 686171 [199]

The furnace here was said to be working in 1653 and discontinued but restocked in 1664. In 1677 the works was leased to Thomas Westerne, a London ironmonger, for five years, but in 1680 the trustees of William Ashburnham's will acquired the works for John Ashburnham and entered into a collusive Chancery suit with Westerne. The Ashburnhams thus came into possession and, although a new lease was granted to Westerne in 1683 for six years, the works appears to have been carried on for the direct benefit of the estate. Early in the eighteenth century they were leased to the Crowley–Hanbury partnership, who remained in occupation until the 1740s (Flinn 1962, 101).

In 1717 Ashburnham was said to produce 350 tons of pig yearly, nearly twice as much as any other Sussex furnace. By 1787 output had dropped to 200 tons. The furnace is one of two Wealden sites (the other being Heathfield) included in the 1794 list, when it was said to be occupied by Lord Ashburnham and to consist simply of a single charcoal furnace. Output in 1796 was given as 173 tons, presumably the actual production, but the furnace was only in use for short periods after 1760 and from 1763 to 1778 in every alternate year. The final date of closure was 1813 (Beswick *et al.* 1984), although the furnace fails to appear in the surveys of 1806 and 1810 (Straker 1931, 364–72; Schubert 1957, 366; Cleere and Crossley 1985, 310–11). The site is a scheduled monument.

Barden, Kent TQ 548425 [188]

Listed in 1653 as working and in 1664 as ruined, although stock accounts for the latter year suggest otherwise, Barden was in use in 1683 and in 1717 had an output of 100 tons. It was still in use in 1729 and in 1761 was being run by William Bowen, who also had Cowden furnace (q.v.) (Hodgkinson 1982, 31-3). Barden appears in the list of closures between 1750 and 1788 as 'Bardam', near Tonbridge, said to be 'down' (Straker 1931, 219; Schubert 1957, 367; Cleere and Crossley 1985, 311–12).

Beckley, Sussex TQ 836212 [199]

In 1664 the furnace was said to be discontinued but restocked. It had been in use in 1653, operated with a forge by Peter Farnden, an arrangement that came to an end with Farnden's death in 1681. In 1715 Samuel Gott was operating the furnace, which had an output of 200 tons in the 1717 list. In 1722 Gott left his son the furnace, which is marked on Budgen's map of 1724 and in the 1740s was being run by the Harrison partnership. Beckley appears in the list of closures between 1750 and 1788 as 'standing', with the owner named as Miss Gott (Straker 1931, 348–9; Schubert 1957, 367; Cleere and Crossley 1985, 313).

Bedgebury, Kent TQ 739347 [188]

The list of 1667 notes that Bedgebury had been discontinued before 1664 but restocked for the second Dutch War; George Browne, the gunfounder, was in partnership with the Foleys in 1657 and was casting guns there in 1665 and 1673–7. It has no later history; the site is a scheduled monument (Straker 1931, 282; Schubert 1957, 367; Cleere and Crossley 1985, 314).

Beech, Sussex TQ 728167 [199]

In use in 1653; discontinued but restocked in 1664. In 1671 William Hawes held 'Beechers' furnace, which may be this site. In 1708 it was occupied by Maximilian Westerne; in 1717 output was recorded as 120 tons; in 1724 the furnace was marked on Budgen's map of 1724. In the latter year Richard Hay of Battle leased the furnace to Lord Ashburnham and Sir Thomas Webster for nine years. It probably did not work after 1737, when Webster leased out Robertsbridge and undertook not to use Beech as a furnace (Straker 1931, 325–6; Schubert 1957, 367–8; Cleere and Crossley 1985, 314; Whittick 1992, 34-5).

Brede, Sussex TQ 801192 [199]

Peter Farnden, in partnership with Samuel Gott, leased the works to Thomas Westerne and Charles Harvey in 1660; it was said to be at work in 1663 and 1664. It was sold to the Westerne family in 1693. In 1717 the output was 200 tons and the site appears on Budgen's map of 1724. William Harrison was casting guns at Brede in 1735 and his operations are well documented in the period 1741–7. The furnace closed in 1766 when the site became a gunpowder mill (Straker 1931, 341–4; Schubert 1957, 368–9; Cleere and Crossley 1985, 318).

Burningfold, Surrey TQ 004343 [186]

A furnace and forge at Burningfold in the parish of Dunsfold are mentioned in 1656 and 1657; in 1673 John Aubrey refers to an 'iron mill'. The works may have survived into the period in which J. Richardson was rector of Dunsfold (1680–1742), although according to Straker the pond had become heath and furze by 1722 (Straker 1931, 422–3; Schubert 1957, 369; Cleere and Crossley 1985, 320–1). J.S. Hodgkinson (pers. comm.) suggests that this site, rather than Fernhurst (q.v.) as previously suggested, may be the 'Burhamfold' furnace, operated by John Butler, mentioned in the list of closures between 1750 and 1788. A militia return of 1758 mentions a 'furnaceman' in the parish (*VCH Surrey*, III.92)

Coushopley, Sussex TQ 604302 [188]

Listed as working in 1653 and 1664 but nothing further is known until 1692, when it was in the hands of the Penkherst family. In 1712 the site was known as 'Pothouse'; it appears in the 1717 list but with no output shown (Straker 1931, 288; Schubert 1957, 372; Cleere and Crossley 1985, 324).

Cowden, Kent TQ 454399 [188]

Recent research (Hodgkinson 1993) has shown that a 'Lower' Cowden furnace mentioned by previous writers (Cleere and Crossley 1985, 325) probably did not in fact exist; no furnace at Cowden appears in 1717 but a map of 1748 shows Upper Cowden complete and apparently in use by William Bowen, a gunfounder mentioned in the Fuller letters between 1747 and 1764.

Crowhurst, Sussex TQ 757122 [199]

In 1627 Peter Farnden leased both forge and furnace at Crowhurst, keeping them until 1653, when Samuel Gott was named as tenant. The furnace was listed in 1653 and 1664 but the forge, in use in 1653, was out of action in 1664. It appears to have no later history (Straker 1931, 352; Schubert 1957, 372; Cleere and Crossley 1985, 326–7).

Darwell, Sussex TQ 701280 [199]

Probably built in 1649, when the Rev. John Gyles obtained licence from the Earl of Winchilsea and others to take earth to make and repair bays. On his death it passed to Benjamin Scarlett. It was advertised for sale in 1694, when it was described as 'lately built' (or perhaps rebuilt). In 1711 John Nicoll of Hendon Place mentioned in his will the manor of Mountfield and his iron furnace and forge. The output in 1717 was 150 tons and the furnace appears on Budgen's map of 1724; in the 1730s the executors of John Crowley had a lease of Darwell, together with Ashburnham (Flinn 1962, 101). The site is mentioned by Fuller in 1737 and in a letter of 1777 is said to have been occupied by its owner and his tenants until the last peace with France and Spain, although it was now in decay and would probably not be revived.

This points to 1763 as the date of closure; in the list of 1750–88 Darwell is coupled with Robertsbridge (q.v.) and the operator of both given as 'Bourne', with the comment 'down' (Straker 1931, 308–9; Schubert 1957, 373; Cleere and Crossley 1985, 327–8).

Ewhurst, Sussex TQ 810248 [199]

Listed as working in 1653 and discontinued but restocked in 1664, this site appears to have no other history apart from a reference to tools at Ewhurst in 1664; it lies close to Northiam (q.v.), with which there may have been some confusion (Straker 1931, 320; Schubert 1957, 375; Cleere and Crossley 1985, 330–1).

Fernhurst, Sussex SU 878283 [186]

The furnace at Fernhurst, known as North Park, was built in 1614, was operating in 1653 but was ruined in 1664. A map of 1660 shows a sketch of the furnace. The site appears to have been revived about 1762 by John Butler as a gun foundry. After he had apparently given up the works it was taken (in 1769) by Wright & Prickett, gunfounders of Southwark, who were succeeded by James Goodyer, a Guildford ironmonger, who was bankrupt in 1777 (Magilton 1990, 33-5; cf. Straker 1931, 426–7; Cleere and Crossley 1985, 331). The site which appears in the list of closures between 1750 and 1788 as 'Burhamfold', operated by Butler and said to be 'down', is probably Burningfold (q.v.), rather than Fernhurst, as previously suggested. Recent investigations at the latter site have revealed remains of the sixteenth-century pond bay (Egan 1990, 205, where the two halves of the grid reference are transposed).

Frith, Sussex SU 955309 [186]

Said to be working in 1653 and 1664, omitted from the list of 1717, but marked on Budgen's map of 1724. A suggested date of closure of 1776 results from confusion with Fernhurst; Frith probably did not work after 1700 (J.S. Hodgkinson, pers. comm.; cf. Straker 1931, 428; Schubert 1957, 375; Cleere and Crossley 1985, 332–3). The site is a scheduled monument.

Gloucester: see Lamberhurst

Gravetye, Sussex
TQ 366342 [187]

Apart from a unsubstantiated reference in 1574, nothing appears to be known of this site until the 1760s, when it was operated by Clutton & Co. William Clutton was bankrupt in 1762 and the same year guns were carried from Gravetye for Eade & Wilton, but in the following year Ralph Clutton and Samuel Durrant were the consignors. In 1768 the occupiers were Raby & Rogers. The furnace appears in the list of closures between 1750 and 1788 (but not in any earlier list) with the comment 'down' and the name of the operators given as Raby & Co. Gravetye was worked at this date in conjunction with Warren furnace (q.v.) (Straker 1931, 236; Schubert 1957, 376; Cleere and Crossley 1985, 333).

Hamsell, Sussex
TQ 538344 [188]

Appears in the lists of 1653 and 1667; in 1674 John Baker sold metal made for shot at Hamsell to George Browne. Baker leased the furnace to John Browne in 1677 for casting ordnance. Robert Baker was bankrupt in 1708, and this may have marked the end of the furnace, which is not listed in 1717. On the other hand, William Harrison used Hamsell for casting ordnance between 1744 and 1750 and an air furnace was built there in 1745. The 1750–88 list gives the occupier as Messrs Harrison (as at Lamberhurst) and describes the furnace as 'down' (Straker 1931, 262; Schubert 1957, 376; Cleere and Crossley 1985, 333–4).

Hawkhurst, Kent
TQ 774313 [188]

A furnace here is first referred to in 1644 and by 1657 was being used by the Foley–Courthope–Browne partnership. In 1660 the Foley share was sold to George Browne and Alexander Courthope and there is an inventory of 1664 of Browne's goods at both the furnace and forge. The same year the furnace was listed as stocked but out of use. In 1668 John Browne entered the partnership. The furnace was listed in 1717 but with no output; it does not appear to have any later history (Straker 1931, 321–2; Schubert 1957, 377; Cleere and Crossley 1985, 334).

Heathfield, Sussex   TQ 599187 [199]

The furnace here was the centre of the Fuller family's activities in the eighteenth century; although there are earlier references to a furnace at Heathfield, the Fullers' works appears to have been newly built in 1693. In 1717 the furnace had an output of 200 tons and in 1724 is marked on Budgen's map. In 1787 it was said to be casting 100 tons annually and appears in the 1794 list as being owned by Mr J. Fuller and consisting simply of a single charcoal furnace, although Cleere and Crossley say that it was last used a year earlier (Straker 1931, 374–6; Schubert 1957, 377; Cleere and Crossley 1985, 335; Crossley and Saville 1991).

Horsmonden, Kent   TQ 695412 [188]

The furnace here was taken over by the Browne family at the beginning of the seventeenth century, their involvement lasting until 1668, during which period it was one of the most important sites in the Weald. It was listed as working in 1667. Any later history is obscure: it does not appear in the 1717 list, but in 1668 a partnership was established between George and John Browne, Alexander Courthope of Horsmonden and William Dyke of Frant to work this and Hawkhurst furnace for thirteen years; according to Straker, Horsmonden was still working in 1689. The Harrisons had a boring-mill in the parish in 1744 but it is not clear whether this was on the same site as the furnace (Straker 1931, 280–1; Schubert 1957, 378; Cleere and Crossley 1985, 337).

Horsted Keynes, Sussex   TQ 379287 [187]

The furnace was said to be in use in 1653 and in repair in 1664; it is mentioned in the will of Edward Lightmaker (1658), in which it was left to his wife Saphirah. A local diary mentions 'men at the furnace' in 1668, which seems to the last reference to it (Straker 1931, 410–11; Schubert 1957, 378; Cleere and Crossley 1985, 337–8).

Imbhams, Surrey
SU 932329 [186]

This site was in use in 1653 and was said to be equipped to make guns in 1664. In the latter year George Browne and Alexander Courthope leased it from John Yalden and stocked it. In 1666 the problem of carrying heavy guns from Imbhams through Guildford is mentioned in a letter. The 1667 list shows that it was laid aside by Browne and the site appears to have no later history (Straker 1931, 420–1; Schubert 1957, 378; Cleere and Crossley 1985, 338–9).

Iridge, Sussex
TQ 749277 [199]

According to Straker, followed by Schubert, a furnace at Iridge is marked on maps of 1710 and 1751. It does not appear in the lists of 1653, 1664 or 1717 and the only reference noted by Cleere and Crossley was in a settlement of 1654 (Straker 1931, 320; Schubert 1957, 379; Cleere and Crossley 1985, 339).

Lamberhurst, Kent
TQ 662360 [188]

The furnace here, sometimes called Gloucester, was built by William Benge in 1695. Samuel Gott was the owner from soon after 1700. In 1717 output was 200 tons; in 1734 Swedenborg described and sketched the furnace, which is marked on Budgen's map of 1724. John Legas and Thomas Hussey leased the furnace in the early 1720s and in 1743 Legas brought the furnace into a group of works run in partnership with William Harrison. In 1782 Hasted mentioned Lamberhurst as the only foundry in Kent; five years later Weale described it as 'yet standing, and possibly may work again in case of war'. In 1784 Lamberhurst was in the hands of William Collins and George Matthews and ore was being bought from a local farmer. The 1750–88 list names Messrs Harrison as operators (as at Hamsell) and makes the same comment: 'stands' (Straker 1931, 269–73; Schubert 1957, 379; Cleere and Crossley 1985, 340–1; Melling 1961, 95).

Mayfield, Sussex   TQ 593282 [199]

The furnace was listed as working in 1653 and repaired in 1664. It appears to have no later history and is referred to as an 'old furnace' in 1712. It is not in any of the eighteenth-century lists. The site is a scheduled monument (Straker 1931, 292–2; Schubert 1957, 381; Cleere and Crossley 1985, 344).

Mill Place, Sussex   TQ 374349 [187]

The furnace here was in use in 1653, discontinued in 1664 but then restocked. It is not in the 1717 list but in 1763 Robert Knight carried 100 guns for Clutton & Durrant, who then owned Mill Place and operated Gravetye; it has not been established that Mill Place was then in blast (Straker 1931, 236–7; Schubert 1957, 382; Cleere and Crossley 1985, 345).

Northiam, Sussex   TQ 817245 [199]

Cleere and Crossley (1985, 347) note that Northiam furnace was in use in 1653, was out of use in 1664 but was subsequently restocked, although it does not appear to have any later history. Their reference to the corresponding entry in Straker (1931, 320), followed by Schubert (1957, 375), leads to entries in those gazetteers under Ewhurst (q.v.), where there was another furnace.

North Park: see Fernhurst

Old Forge, Sussex   TQ 459258 [198]

This was an important sixteenth-century furnace rebuilt between 1614 and 1619. It has no later history, apart from a reference in 1664 to a furnace at Maresfield 'continued in hope', which may or may not have been this one (Straker 1931, 398–9; Schubert 1957, 381; Cleere and Crossley 1985, 347).

Pallingham, Sussex TQ 041227 [197]

The furnace here was working in 1653 and 1664 but apparently has no later history (Straker 1931, 425; Schubert 1957, 383; Cleere and Crossley 1985, 348).

Pippingford, Sussex TQ 450316 [187]

A furnace here was evidently built between 1693, when it fails to appear in a survey, and 1696, when this part of Ashdown Forest was enclosed. In 1717 Charles Hooper leased the furnace to Charles Manning of Dartford, but whether Manning had a previous tenancy is not clear. The site appears as 'New Furnace' on Budgen's map of 1724 but no ironworks is marked on a map of 1738. The 1717 list includes an entry for 'Mr Pounsley Ashdowns Forrest', with no output shown; it is not entirely clear whether this refers to one or two sites (cf. Pounsley). The surviving remains are a scheduled monument (Straker 1931, 247–8; Cleere and Crossley 1985, 350).

Pounsley, Sussex TQ 529219 [199]

The furnace here, which is a scheduled monument, was working in 1653 and in 1664, although discontinued, was restocked. In 1671 ordnance was being transported to South Malling from Framfield, probably originating from Pounsley. It is marked on Budgen's map of 1724 but the entry in the 1717 list (see Pippingford) is ambiguous: 'Mr Pounsley' may be an error for this site or may refer to Pippingford; in any case, no output is shown (Straker 1931, 391; Schubert 1957, 384; Cleere and Crossley 1985, 351).

Robertsbridge, Sussex TQ 751231 [199]

The furnace here was in use in 1653 and 1667. In 1680 John Roberts held it for the Earl of Leicester and in 1707 John Snepp, senior and junior, were the occupiers. The furnace was producing 120 tons in 1717. Sir Thomas Webster bought the estate in 1721 and from 1724 to 1734 carried on the furnace and forge himself. In the latter year he made a lease to Harrison, Jukes (or Jewkes) & Co. of London, and granted another to William and George Jukes for seven years from December 1746. In 1754 the works was taken by John Churchill of Hints (Staffs.) for a further term of seven years; correspondence between Churchill and Sir Whistler Webster, the landowner, makes it clear that it was the Jukes brothers who built an air furnace at Robertsbridge, not Churchill, as previously suggested (Whittick 1992, 34-62). In 1768, after Churchill's bankruptcy and death (cf. Rushall, Staffs.) Robertsbridge was leased to William Polhill of Hastings, David Guy of Rye and James Bourne of Salehurst, ironmasters. Bourne alone was said to be the occupier in 1785, as he is in the list of closures between 1750 and 1788 (as at Darwell, q.v.), when the furnace was 'down'. Weale's version says Robertsbridge was then 'standing', ready to work in case of war, and could produce 50 tons a year. The furnace was last assessed to local rates in 1793 and the forge was sold in bankruptcy in 1801 (Straker 1931, 310–18; Schubert 1957, 385; Cleere and Crossley 1985, 352–3).

Scarletts, Kent　　　　　　　　　　　　　　　　　TQ 443401 [187]

Stock was listed here in 1655–7 and the furnace was in use in 1664, making guns or shot for the Dutch wars. Its later history is difficult to separate from that of the nearby Cowden furnace (q.v.), although Straker showed reasonably clearly that the furnace mentioned by Leonard Gale in 1670 and later acquired by his son Leonard (either by purchase or marriage with Sarah Knight in 1703) was Scarletts, since after his death in 1750 a private Act of 1761 included the site on which Scarletts furnace had stood. It was out of use by this date and is not mentioned in the list of 1717 (Straker 1931, 224–7; Schubert 386; Cleere and Crossley 1985, 354–5).

Socknersh, Sussex　　　　　　　　　　　　　　　TQ 705233 [199]

The furnace here was working in 1653, then went out of use but was restocked in 1664 for the war. In 1671 it was leased by Thomas Collins for four years to Peter Farnden and John Roberts. It was not listed in 1717 and although marked on Budgen's map of 1724 may have been out of use by this date (Straker 1931, 306–7; Schubert 1957, 387; Cleere and Crossley 1985, 357).

Sowley, Hants. SZ 3796 [196]

The furnace and forge here were built about 1605 by the Earl of Southampton and was still in use in 1717, when the output was given as 200 tons. It was then the only furnace in Hampshire. In the list of closures between 1750 and 1788 there is an entry under Sussex which reads 'Sowley Hants ... Fold ... down', and another at the end of the list: 'Sawley ... Todd ... down', both apparently referring to this site (Schubert 1957, 388). The Knights' Stour Partnership accounts record purchases of pig from Sowley in 1734–8 (when the ironmaster's name was given as Hall), 1767–71 (William Ford) and 1771–3 (John Dixon) (Ince 1991, 118).

Stream, Sussex TQ 555155 [199]

The furnace and forge here were mentioned in 1653 and 1667. Guns were cast at Stream in 1692–3 by the Fullers, before they built the furnace at Heathfield. Stream is not listed in 1717 but appears on Budgen's map of 1724 (Straker 1931, 384; Schubert 1957, 388; Cleere and Crossley 1985, 194, 359).

Tilgate, Sussex TQ 284355 [187]

The furnace here is first mentioned in 1574 and was in use in the first half of the seventeenth century. It was noted as discontinued but repaired and restocked on account of the war in 1664 but appears to have no later history (Straker 1931, 465; Schubert 1957, 389; Cleere and Crossley 1985, 361).

Vauxhall, Kent TQ 593440 [188]

There are leases for this site from 1622 to 1679 but it is not clear for how much of that period the furnace was in use. It does not appear in the lists of 1653 and 1664, although, according to Schubert, both furnace and forge were in use in 1670. They were described as 'now demolished' in 1701 (Straker 1931, 222; Schubert 1957, 388; Cleere and Crossley 1985, 362).

Waldron, Sussex TQ 566181 [199]

The operation of the furnace here by the Pelhams is well documented for most of the period 1639–1715. The site appears in the lists of 1653 and 1664–7. In 1717 it was making 150 tons and is marked on Budgen's map of 1724. The furnace made pig and shot rather than ordnance in the seventeenth century but produced guns from 1747, by which time it was run within the partnership that included Legas and Harrison. According to Straker, the site was acquired by the Fullers in 1716, which corresponds with the evidence of the 1750–88 list, in which Waldron is said to be 'down', with the name J. Fuller alongside (as owner, there is no evidence that he ever worked the furnace; ex inf. J.S. Hodgkinson). The exact date of closure is not known (Straker 1931, 381–2; Schubert 1957, 390; Cleere and Crossley 1985, 363).

Warren, Sussex TQ 348393 [187]

The history of this site falls into two distinct periods, one from before 1567 to about 1627; and then a revival in the late eighteenth century, when Edward Raby cast iron and probably also brass ordnance here from about 1762 to 1774. Straker describes the later occupiers as 'Masters & Raby', latterly 'Raby & Rogers'. The 1750–88 list gives the occupiers as Raby & Co. (as at Gravetye, q.v.), and describes the furnace as 'down'; the site is a scheduled monument (Straker 1931, 214–7; Schubert 1957, 390; Cleere and Crossley 1985, 364).

Warsash, Hants. SU 4906 [196]

According to *Mineral Statistics*, a charcoal-fired furnace at Warsash, near Southampton, was under construction in 1868 and was put into blast in 1869. It was then out of blast until 1874 and worked again (for the last time) in 1877. The furnace was built and operated by Harrison Ainslie & Co., the Furness ironmasters who were still using charcoal at Backbarrow and Newland, Lancs. (qq.v.); the site at Warsash appears to have the distinction of being the last furnace to be built anywhere in Great Britain to smelt iron ore with charcoal (although not, of course, the last to close down). The works are described in a local publication of c. 1940: F.W.L., *A short history of Warsash* (pp. 23–6).

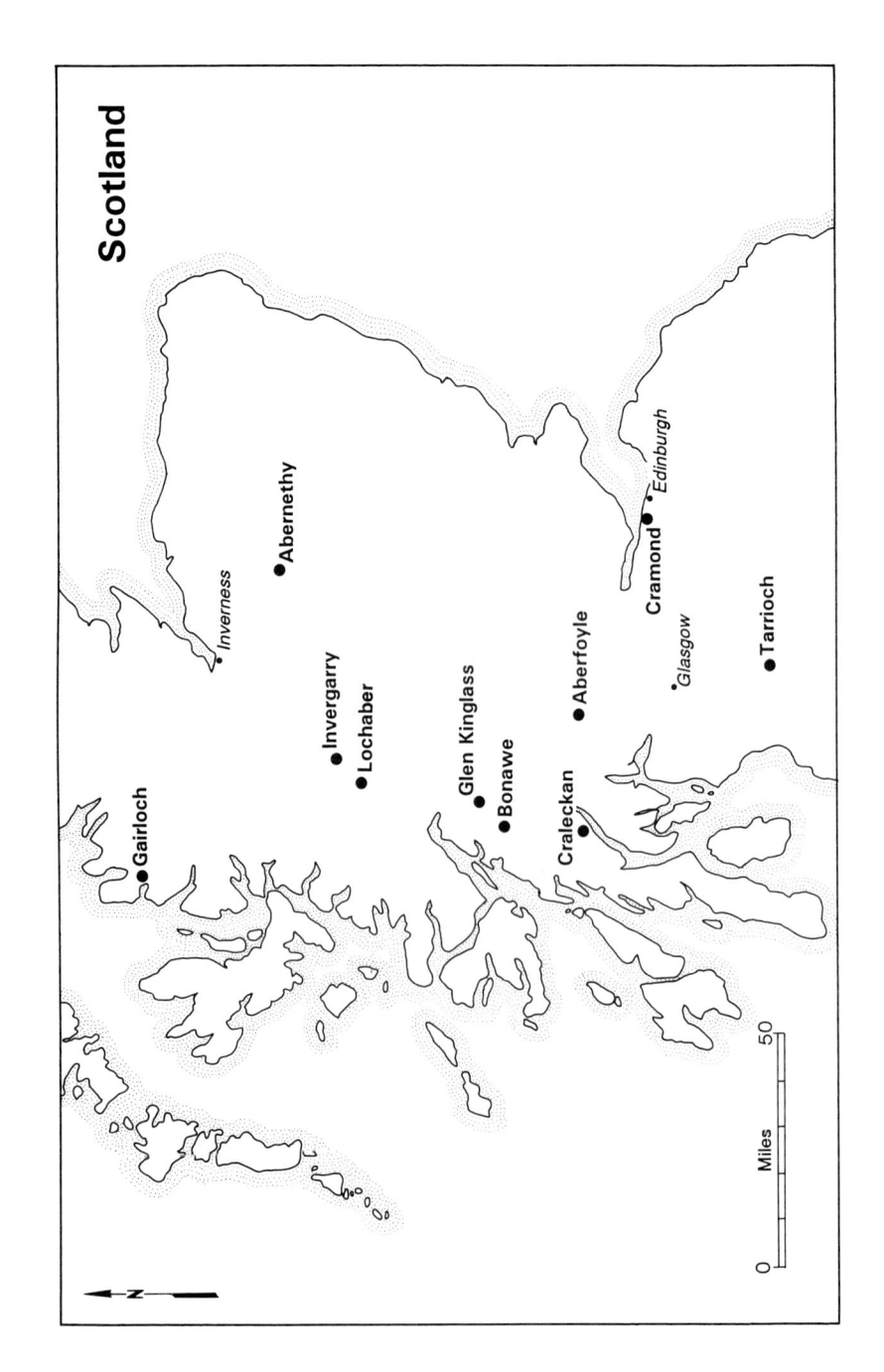

12

Scotland

Schubert's history of the 'British' iron industry devoted half a page to charcoal blast furnaces in Scotland (1957, 191) and omitted them entirely from the gazetteer. More recently J.M. Lindsay (1977) and J.H. Lewis (1984) have rescued the subject from the neglect it had suffered since W.I. Macadam's well known article of 1887 and the entries here are based largely on their work.

Aberfoyle, Perthshire NN 5200 [57]

In April 1718 John Smith of Castlefinn, Co. Donegal, and John Irvine of Newtonwood, Co. Tyrone, bought the right to cut wood, prospect for ore and build an iron-mill from the Montrose estate at Buchanan and Menteith. The works was in existence by June 1722, under the control of a partnership which including John Gordon of Kirkconnel, James Graham of Kilmannan and several Glasgow merchants. Known as Graham's Miln, it was situated near Achray, to the east of Loch Katrine, and seems to have comprised a forge only, without a furnace, although there is one inconclusive reference to ore being used. Otherwise the mill reworked imported scrap. It had a short life, possibly shutting as early as 1724 (Lindsay 1977, 55).

Abernethy, Inverness-shire NH 996217 [36]

A furnace, known either as Abernethy or Culnakyle, was built near the confluence of the Spey and its tributary the Nethy in Strathspey by the York Buildings Co., probably in 1729. There was also a forge but the works were unsuccessful and probably shut in 1734 or 1735. This was one of several Scottish ventures by the same company which collapsed around this date; Abernethy was more than 50 miles from the nearest port and ill-placed to serve any major market (Lindsay 1977, 59–60; J.H. Lewis 1984, 44–5).

SCOTLAND

Bonawe, viewed from the north-west, after excavation and consolidation, with the blowing-arch on the left and the tapping-arch on the right. The water-wheel was beyond the far wall of the blowing-house (*Crown Copyright: Historic Scotland*).

Argyll: see Craleckan

Bonawe, Argyllshire NN 009318 [50]

The furnace at Bonawe on Loch Etive, also known as Lorn, was built by Richard Ford & Co., i.e. the Newland Co., which, as Harrison Ainslie, eventually secured control of the whole of the Furness industry. The surviving building is dated 1753 and contracts with landowners were made the following year, ensuring a long-term supply of charcoal. These agreements expired only in 1863, after which the furnace worked intermittently until 1876. The ore came mostly from Furness and coastal shipping brought charcoal from a wide area. The furnace was capable of making 700 tons of pig annually, most of which was sent to north-west England, the Severn estuary and South Wales. Cheap fuel gave it an advantage over other Highland furnaces which probably contributed to its long life. Only Backbarrow and Newland (Lancs.) (qq.v.) outlived Bonawe; all three were worked by Harrison Ainslie in the nineteenth century (Lindsay 1975; 1977, 60–1; J.H. Lewis 1984, 466–73).

Despite its northerly location, Bonawe had a long-standing connection with the Knights' Stour Partnership forges, supplying pig almost every year from 1755 until 1810. Between 1755 and 1764 the tenant at the furnace was named in the accounts as William Ford; in 1771–6 as John Dixon; in 1776–85 as George Knott and thereafter until 1800 as his executors; and finally, between 1805 and 1810, as George Buckle (Ince 1991, 117).

In 1796 the furnace was listed as 'Bower' and coupled with the other surviving Scottish charcoal furnace at Craleckan (q.v.), the two together being ascribed an output of 1,600 tons a year, which should probably be interpreted as 16 tons a week each. In the 1794 list the furnace appear as 'Bunawe', the owner as the Duke of Argyll and the occupier as L.G. Knott. There was a single furnace, blown by water, to which the compiler assigned a date of building of '174–'. In 1806 output at Bonawe was said to be 400 tons p.a.; Truran (1855) suggested a figure of 800 tons. The furnace was listed (as Lorn) in *Mineral Statistics* from 1855 until 1881, although it was only in blast in 1855–8, 1863, 1868–70 and 1873–6. The returns noted that it was supplied with haematite from Furness and used charcoal as fuel.

The substantially complete remains of the works at Bonawe are now in the guardianship of the Secretary of State for Scotland, whose officials have published two exceptionally good descriptions of the site (Stell and Hay 1984; Hay and Stell 1986, 108–14; cf. also Butt 1967, 106–7). Bonawe thus

Craleckan, viewed from the north-east, with the casting-house in the foreground and the ruined blowing-house with charging-house over beyond to the right. *(Crown Copyright: Historic Scotland)*

ranks alongside Duddon (Lancs.) and Dyfi (Cards.) (qq.v.) as one of three best preserved and best presented charcoal blast furnaces in Britain. Besides the furnace itself, the remains at the working area including a complete, roofed charging-house and the consolidated ruins of the blowing-house and casting-house. On the higher ground to the south of the furnace stand three large charcoal stores and an iron ore store. Outside the area in the guardianship of the Secretary of State housing built for workers at the furnace can be seen; Bonawe House, the manager's home, stands in private wooded grounds a little further away. To complete a visit to this exceptionally interesting and spectacularly situated site it is worth seeing the masonry pier which extends some 130m. into the waters of Loch Etive about 300m. north-west of the furnace. It was here that iron ore from Furness was delivered and pig loaded for England.

Craleckan, Argyllshire NM 025001 [55]

The furnace known variously as Argyle, Craleckan or Goatfield, standing in the village of Furnace on Loch Fyne, was built by Jonathan Kendall & Co., generally known as the Duddon Co. from their main base of operations in Furness. Contracts for wood date from 1754 and a furnace was under construction the following year. The surviving inscription on a lintel, '1775 CF', apparently refers to a rebuilding, not original establishment, since shipments of ore were arriving by 1757, if not earlier. For a short period (1760–3) Kendall & Co. were selling pig from Craleckan to the Knights' Stour Partnership (Ince 1991, 117). The furnace was blown out in 1813; in later years there was a forge as well as a casting-house on the site (Lindsay 1977, 61; Butt 1967, 107–8).

In the 1796 list Argyll is coupled with 'Bower' (i.e. Bonawe, q.v.) the two together said to make 1,600 tons a year, with charcoal as the fuel, presumably meaning 16 tons a week from each furnace. In the 1794 list the furnace appears with the owner as the Duke of Argyle and the occupier Messrs Kendall. There was one charcoal furnace, blown by water, said to have been built in 1755. The 1806 survey lists two furnaces at the site (in obvious error, since Bonawe has its own entry), both out of blast, with no output shown and no owner named.

At the site today, which is privately owned but can be inspected from the minor road off the A83 which serves the village of Furnace, it is possible to see a substantially complete furnace stack, with the ruins of the charging-house to the rear and other ancillary buildings (cf. Butt 1967, 107–8, Lewis 1984, 473–6, Hay and Stell 1986, 114–15).

Cramond, Midlothian NT 1876 [65]

A furnace at Cramond, five miles west of Edinburgh, appears in the 1794 list as being owned and occupied by Cadell & Eddington, with the plant consisting of a single charcoal furnace (the means of blowing and date of building not stated) and a slitting-mill, built in 1760. According to Butt (1967, 109–10) the works were converted from four corn-mills, as a forge only, the first lease dating from 1752. Cramond at first relied on imported Swedish and Russian bar, which was presumably slit or reworked in other ways there. In 1759 Cramond was bought by the Carron Co., which had established the first coke furnace in Scotland, but this union lasted only until 1770. It was then bought by the Cadells, who continued to own it until 1860. Neither a detailed history of the site (Cadell 1973) nor Butt suggests that smelting ever took place at Cramond; Cadell specifically states that the forge refined bar from elsewhere. No furnace is listed here in 1796 or 1806, which possibly suggests that the 1794 list is (for once) in error.

Culnakyle: see Abernethy

Gairloch, Ross & Cromarty NG 8076 [19]

Lindsay (1977, 49–52) assembled scattered evidence for attempts to smelt iron on three sites in the parish of Gairloch in Ross & Cromarty between 1607 and 1634. Little is known of any of these ventures and there is no evidence that they were in use after 1660.

Glen Kinglass, Argyllshire NN 082371 [50]

A furnace near the mouth of the River Kinglass in Argyllshire, close to Loch Etive, is first mentioned in 1725. It was probably established sometime after May 1722, the date of the first deed of copartnery. In 1725 the partners were Capt. Arthur Galbraith of Dublin, Roger Murphey, tanner, of Enniskillen, Charles Armstrong of Mountarmstrong, Kildare, and William Ketlewell of

Thomastown, Meath. Little is known of the works, except that it was undoubtedly a blast furnace, since castings were supplied to the York Buildings Co. leadworks at Strontian around 1730. Charcoal was probably the main fuel, although Macadam claimed that peat was tried and rejected. As at other works in the region, ore was imported from Furness. The date of closure is uncertain; it was certainly defunct by 1752 and may have shut in the depression of 1737–8 (Lindsay 1977, 56–7). J.H. Lewis (1984, 445–64) describes the site and the results of his excavations there (cf. also Hay and Stell 1986, 108).

Goatfield: see Craleckan

Invergarry, Inverness-shire　　　　　　　　　　　　　　NH 313010 [34]

The short-lived enterprise at Invergarry originated in a contract of 1727 by which a Lancashire partnership headed by Thomas Rawlinson bought woods from John Macdonald of Invergarry. Construction started soon afterwards but the furnace was not blown in until 1729. Production was then intermittent; nothing was made for twelve months after February 1734 and the furnace was finally blown out in early in 1736. Most of the metal was sold as pig, although a few castings were made for Strontian leadworks in 1730. The pig mainly went to Chepstow, where it was sold by Nehemiah Champion and later John Beckett. Invergarry was too far inland to compete successfully in the English market; the main locational incentive was availability of fuel and the ore, as at other sites in the region, was imported from Lancashire (Lindsay 1977, 58–9; Butt 1967, 105; J.H. Lewis 1984, 445).

Lochaber, Inverness-shire　　　　　　　　　　　　　　NN 1788? [34]

Lindsay (1977, 54–5) notes a reference in 1688 to iron being made at a mill on the River Arkaig in Lochaber, Inverness-shire, managed by John Davidson. It appears to have no other history and probably did not have a long life on either side of 1688.

Lorn: see Bonawe

Tarrioch, Ayrshire
NS 642269 [71]

J.H. Lewis (1984, 439, 463–5) appears to be the first modern writer to notice what was apparently the only charcoal blast furnace in south-west Scotland, at Tarrioch (or Terrioch) near Muirkirk. It is thought to have been built about 1732 by the then Lord Cathcart, to have used haematite mined nearby and to have had an associated forge. The furnace is said to have been abandoned at an early (though as yet undetermined) date when local charcoal supplies were exhausted and attempts to smelt with peat were unsuccessful. The furnace does not appear in any of the nationally compiled lists. Its existence, however, has been confirmed by fieldwork by Lewis, who located a collapsed furnace stack and other remains.

Bibliography

The place of publication is London unless otherwise stated. For details of county maps referred to in the text see Elizabeth M. Rodger, *The large scale county maps of the British Isles 1596–1850. A union list* (Bodleian Library, 2nd ed., 1972).

'An early Neath Abbey works', *Trans. Neath Antiquarian Soc.*, 2nd series, 3 (1932–3), 82–3.

'Miscellanea. Local partnership deeds. (From the MS. collection of Mr Richard Welford.)', *Proc. Soc. of Antiquaries of Newcastle-upon-Tyne*, 3rd series, 3 (1907–8), 170–1.

Aikin, J., *A description of the country from thirty to forty miles round Manchester* (1795).

Allen, E., *see* Raistrick, A.

Andrews, C.R., *The story of Wortley ironworks* (2nd ed., Worksop, 1956).

Anstis, R., *The story of Parkend, a Forest of Dean village* (Coleford, 1982).

Ashmore, B.G., *see* Tylecote, R.F.

Ashton, T.S., *Iron and steel in the industrial revolution* (Manchester, 1924).

Ashurst, D., *see* Crossley, D.W.

Atkinson, F., *The industrial archaeology of north-east England (the counties of Northumberland and Durham and the Cleveland district of Yorkshire)* (Newton Abbot, 1974).

Awty, B.G., 'Charcoal ironmasters of Cheshire and Lancashire, 1600–1785', *Trans. Historic Soc. of Lancs. & Cheshire*, 109 (1957), 71–124.

Banks, A.P., *see* Tylecote, R.F.

Barnes, F., *Barrow and district. An illustrated history published by the Barrow-in-Furness Library and Museum Committee on the occasion of the Festival of Britain* (Barrow-in-Furness, 1951).

Bayliss, D.G., 'The effect of Bringewood forge and furnace on the landscape of part of northern Herefordshire to the end of the seventeenth century', *Trans. Woolhope Naturalists' Field Club*, 45 (1985–7), 721–9.

Bell, I. Lowthian, 'On the manufacture of iron in connection with the Northumberland and Durham coal-field', in *The industrial resources of the district of the three northern rivers, the Tyne, Wear and Tees, including the reports on local manufactures, read before the British Association in 1863*, ed. Armstrong, W.G., and others (London and Newcastle-upon-Tyne, 1864), pp. 73–119.

Bergen, C., *History of Bedlington ironworks 1736–1867* (Bedlington, n.d. [*c*. 1948]).

Beswick, W.R., Broomhall, P.J., and Bickersteth, J.D., 'Ashburnham blast furnace: a definitive date for its closure', *Sussex Arch. Collections*, 122 (1984), 226–7.

Bick, D.E., 'Remnants of Newent furnace', *Gloucs. Industrial Archaeology Soc. Journal* (1980), 29–37.

'Firebacks', *Period Home* (September 1985), 21–24.

The mines of Newent and Ross (Newent, 1987).

'New light on Blakeney furnace', *New Regard of the Forest of Dean*, 6 (1990), 45–7.

Bickersteth, J.D.: *see* Beswick, W.R.

Blick, C.R., 'Early blast furnace news', *Historical Metallurgy*, 18 (1984), 44–50.

Bradney, J.A., *A history of Monmouthshire from the coming of the Normans into Wales down to the present time* (1904–33).

Broomhall, P.J.: *see* Beswick, W.R.

Brook, F., *The industrial archaeology of the British Isles. 1. The West Midlands* (1977).

Brown, P.J., 'The early industrial complex at Astley, Worcestershire', *Post-Medieval Archaeology*, 16 (1982), 1–19.

Brown, R.R., 'The Woolwich proof registers 1780–1781', *International Journal of Nautical Archaeology & Underwater Exploration*, 17 (1988), 105–111.

Bull, H.G., 'Some account of Bringewood forge and furnace', *Trans. Woolhope Naturalists' Field Club for 1869*, 54–9.

Butt, J., *The industrial archaeology of Scotland* (Newton Abbot, 1967).

Cadell, P., *The iron mills at Cramond* (Edinburgh, 1973).

Cantrill, T.C., and Wight, M., 'Yarranton's Works at Astley', *Trans. Worcs. Arch. Soc.*, new series, 6 (1930), 92–115.

Cave, B.V., 'Mill sites on the Longhope—Flaxley—Westbury streams', *Gloucs. Industrial Archaeology Soc. Journal* (1974), 9–40.

'Gun's Mills, Abenhall', *Gloucs. Industrial Archaeology Soc. Journal* (1981), 2.

Chaloner, W.H., 'Isaac Wilkinson, potfounder', in Pressnell, L.S. (ed.), *Studies in the industrial revolution presented to T.S. Ashton* (1960), 23–51.

Chapman, S.D., *Stanton and Staveley. A business history* (Cambridge, 1981).

Chappell, E.L., *Historic Melingriffith: an account of Pentyrch iron works* (Cardiff, 1940).

Cherry, J., 'Post-Medieval Britain in 1981', *Post-Medieval Archaeology*, 16 (1982), 217–29.

Cleere, H., and Crossley, D., *The iron industry of the Weald* (Leicester, 1985).

Courtney, P., 'The rural landscape of eastern and lower Gwent, *c.* A.D. 1070–1750' (Unpublished University of Wales [Cardiff] Ph.D. thesis, 1983).

'Some new light on the Gwent iron industry in the 17th century', *Monmouthshire Antiquary*, 7 (1991), 65–9.

Courtney, P., and Gray, M., 'The site of Tintern Abbey after the Dissolution', *Bulletin of the Board of Celtic Studies*, 38 (1991), 145–58.

Coxe, W., *An historical tour in Monmouthshire* (1801).
Cranstone, D., 'The iron industry of the Ashby coalfield', *Leics. Industrial History Soc. Bull.*, 8 (1985), 23-31.
— (ed.), *The Moira furnace. A Napoleonic blast furnace in Leicestershire* (N.W. Leics. District Council, 1985).
Crossley, D.W., 'Rockley furnace and engine house', *Archaeological Journal*, 137 (1980), 445-7.
See also Cleere, H.
Crossley, D.W., and Ashurst, D., 'Excavations at Rockley smithies, a water-powered bloomery of the sixteenth and seventeenth centuries', *Post-Medieval Archaeology*, 2 (1968), 10-54.
Crossley, D.W., and Saville, R.V. (ed.), *The Fuller letters 1728-1755. Guns, slaves and finance* (Sussex Record Soc., 76, 1991).

Davies, A.S., 'The charcoal iron industry of Powys Land', *Montgomeryshire Collections*, 46 (1940), 31-66.
— 'The early iron industry in North Wales', *Trans. Newcomen Soc.*, 25 (1945-7), 83-90.
Davies, D., *Hanes plwyf Penderyn* (Aberdâr, 1905).
Davies, D.J., *The economic history of South Wales prior to 1800* (Cardiff, 1933).
Davies, J.H., *History of Pontardawe and district from earliest to modern times* (Llandybie, 1967).
Davies, W., *General view of the agriculture and domestic economy of south Wales*, II (1815).
Davies-Shiel, M., *see* Marshall, J.D.
Dent, R.K., *Old and new Birmingham; a history of the town and its people* (Birmingham, 1880).
Dickinson, H.W., *John Wilkinson, ironmaster* (Ulverston, 1914).
Dinn, J., 'Dyfi Furnace excavations, 1982-87', *Post-Medieval Archaeology*, 22 (1988), 111-42.

Edwards, I., 'The early ironworks of north-west Shropshire', *Trans. Shropshire Arch. Soc.*, 56 (1957-60), 185-201.
— 'The charcoal iron industry of east Denbighshire 1630-90', *Trans. Denbighs. Hist. Soc.*, 9 (1960), 23-53.
— 'The charcoal iron industry of Denbighshire, c. 1690-1770', *Trans. Denbighs. Hist. Soc.*, 10 (1961), 49-97.
— 'Industrial archaeology: North Wales section of the C.B.A.', *Trans. Denbighs. Hist. Soc.*, 13 (1964), 242-3.
— 'Iron production in north Wales: the canal era: 1795-1850', *Trans. Denbighs. Hist. Soc.*, 14 (1965), 141-84.

'John Wilkinson and Boulton & Watt', *Trans. Denbighs. Hist. Soc.*, 21 (1972), 109–116.

'The British Iron Company', *Trans. Denbighs. Hist. Soc.*, 31 (1982), 109–48.

Egan, G., 'Post-Medieval Britain in 1984', *Post-Medieval Archaeology*, 19 (1985), 159–91.

'Post-medieval Britain and Ireland in 1988', *Post-Medieval Archaeology*, 23 (1989), 25–68.

'Post-medieval Britain and Ireland in 1989', *Post-Medieval Archaeology*, 24 (1990), 159–211.

Ellis, H., 'Iron working at Flaxley Abbey', *New Regard of the Forest of Dean*, 1 (1985), 12–17.

Elsas, M. (ed.), *Iron in the making: Dowlais Iron Company letters, 1782–1860* (Glamorgan Record Office, 1960).

England, J., 'The Dowlais iron works, 1759–93', *Morgannwg*, 3 (1959), 41–60.

Evans, C., 'Manufacturing iron in the north-east during the eighteenth century: the case of Bedlington', *Northern History*, 28 (1992), 178–96.

Evans, L.W., 'Robert Morgan of Kidwelly, ironmaster', *Trans. Carmarthenshire Antiquarian Soc.*, 28 (1938), 136–8.

Evans, M.C.S., 'The pioneers of the Carmarthenshire iron industry', *Carmarthenshire Historian*, 4 (1967), 22–40.

'The Llandyfân forges. A study of ironmaking in the upper Loughor valley', *Carmarthenshire Antiquary*, 9 (1973), 131–56.

'Cwmdwyfran forge, 1697–1839', *Carmarthenshire Antiquary*, 11 (1975), 146–76.

Fairclough, O., *The grand old mansion: the Holtes and their successors at Aston Hall 1618–1864* (Birmingham Museums & Art Gallery, 1984).

Farey, J., *A general view of the agriculture and minerals of Derbyshire*, I (1811).

Fell, A., *The early iron industry of Furness and district* (Ulverston, 1908).

Fletcher, H.A., 'The archaeology of the west Cumberland iron trade', *Trans. Cumberland & Westmorland Antiquarian Soc.*, 1st series, 5 (1881), 5–21.

Flinn, M.W., 'Industry and technology in the Derwent valley of Durham and Northumberland in the eighteenth century', *Trans. Newcomen Soc.*, 29 (1955), 255–62.

'William Wood and the coke-smelting process', *Trans. Newcomen Soc.*, 34 (1961–2), 55–71.

Men of iron. The Crowleys in the early iron industry (Edinburgh, 1962).

Galgano, M.J., 'Iron-mining in Restoration Furness: the case of Sir Thomas Preston', *Recusant History*, 13 (1976), 212–18.

Glover, C., and Riden, P. (ed.), *William Woolley's History of Derbyshire* (Derbyshire Record Soc., 6, 1981).

Goodman, K.W.G., 'Hammermen's Hill' (Unpublished University of Keele Ph.D. thesis, 1978).

BIBLIOGRAPHY

'Tilsop furnace', *West Midlands Studies*, 13 (1980), 40–6.

Gray, M., *see* Courtney, P.

Green, F., 'Carmarthen tinworks and its founder', *W. Wales Historical Records*, 5 (1915), 245–70.

Green, H., 'Melin-y-Cwrt Furnace. Earth, air, fire and water', *Trans. Neath Antiquarian Soc.* (1980–1), 43–80.

Griffin, G., 'The home of the Deincourts', *Derbys. Arch. Journal*, 40 (1918), 202–6.

Griffiths, L., Unpublished MS. notes on the history of Dovey and Conway furnaces, 1953? (Welsh Industrial and Maritime Museum, Cardiff).

Hallett, M.M., and Morton, G.R., 'Yarranton's blast furnace at Sharpley Pool, Worcestershire', *Journal of the Iron & Steel Institute*, 206 (1968), 689–92.

Hammersley, G.F., 'The history of the iron industry in the Forest of Dean region, 1562–1660' (Unpublished University of London Ph.D. thesis, 1971).

'The charcoal iron industry and its fuel, 1540–1750', *Economic History Review*, 2nd series, 26 (1973), 593–616.

Harris, F.J.T., 'Guns Mills as a paper mill', *Gloucs. Industrial Arch. Soc. Journal* (1974), 33–39.

Hart, C.E., *The free miners of the royal Forest of Dean and hundred of St Briavels* (Gloucester, 1953).

The industrial history of Dean with an introduction to its industrial archaeology (Newton Abbot, 1971).

Hay, G.D., and Stell, G.P., *Monuments of industry. An illustrated historical record* (Edinburgh, 1986).

See also Stell, G.P.

Hayman, R., *Working iron in Merthyr Tydfil* (Merthyr Tydfil Heritage Trust, 1989).

Hey, D.G., 'The ironworks at Chapeltown', *Trans. Hunter Arch. Soc.*, 10 (1971–77), 252-9.

The making of south Yorkshire (Ashbourne, 1979).

Yorkshire from AD 1000 (1986).

Hodgkinson, J.S., 'The *Sussex Weekly Advertiser* — some extracts', *Wealden Iron*, 2nd series, 2 (1982), 30–6.

'Notes on Kent furnaces', *Wealden Iron*, 2nd series, 13 (1993) (forthcoming).

Hopkinson, G.G., 'Staveley Forge 1762–83', *Trans. Hunter Arch. Soc.*, 7 (1951–7), 94–5.

'A Sheffield business partnership, 1750–65', *Trans. Hunter Arch. Soc.*, 7 (1951–7), 103–17.

'The charcoal iron industry of the Sheffield region, 1500–1775', *Trans. Hunter Arch. Soc.*, 8 (1961), 122–51.

Hughes, E., 'The founding of Maryport', *Trans. Cumberland & Westmorland Arch. and Antiq. Soc.*, 2nd series, 64 (1964), 306–18.

Hughes, S., and Reynolds, P., *A guide to the industrial archaeology of the Swansea region* (Association for Industrial Archaeology, 1988).

Hulme, E.W., 'Statistical history of the iron trade of England and Wales, 1717–50', *Trans. Newcomen Soc.*, 9 (1928–9), 12–35.

Hutchinson, W., *The history and antiquities of the county palatine of Durham* (Newcastle-upon-Tyne, 1787).

Hyde, C.K., 'The iron industry of the West Midlands in 1754: observations from the travel account of Charles Wood', *West Midlands Studies*, 6 (1973), 39–40.

Technological change and the British iron industry 1700–1870 (Princeton, N.J., 1977).

Ince, L., *The Neath Abbey Iron Company* (Eindhoven, 1984).

The Knight family and the British iron industry, 1692–1902 (Birmingham, 1991).

James, T., 'Carmarthen tinplate works 1800–1821', *Carmarthenshire Antiquary*, 12 (1976), 31–54.

Carmarthen: an archaeological and topographical survey (Carmarthen Antiquarian Soc., 1980).

Jenkins, R., 'Ironmaking in the Forest of Dean', *Trans. Newcomen Soc.*, 5 (1925–6), 42–65.

The great Jennens case: being an epitome of the history of the Jennens family. Compiled on behalf of the Jennens family by Messrs Harrison and Willis (Sheffield, 1879).

John, A.H. (ed.), *Minutes relating to Messrs Samuel Walker & Co., Rotherham, iron founders and steel refiners, 1741–1829, and Messrs Walkers, Parker & Co., lead manufacturers, 1788-1893* (Council for the Preservation of Business Archives, 1951).

'Introduction. Glamorgan, 1700–1750', in John, A.H. and Williams, G. (ed.), *Industrial Glamorgan from 1700 to 1970* (Glamorgan County History, 5, 1980).

Johnson, B.L.C., 'The charcoal iron industry in the Midlands, 1690–1720' (Unpublished University of Birmingham M.A. thesis, 1950).

'The Foley partnerships: the iron industry at the end of the charcoal era', *Economic History Review*, 2nd series, 4 (1951–2), 322-40.

'New light on the iron industry in the Forest of Dean', *Trans. Bristol & Gloucs. Arch. Soc.*, 72 (1953), 129–43.

'The iron industry of Cheshire and north Staffordshire, 1688–1712', *Trans. N. Staffs. Field Club*, 88 (1953–4), 32–55.

Johnson, R., '17th century iron works at Bulwell and Kirkby', *Trans. Thoroton Soc. Notts.*, 64 (1960), 44–46.

Jones, D.W., *Hanes Morganwg* (Aberdâr, 1874).

Jones, E., *A history of GKN* (1987–90).

Jones, E.G. (ed.), *Exchequer proceedings (equity) concerning Wales. Henry VIII–Elizabeth. Abstracts of bills and inventory of further proceedings* (Cardiff, 1939).

Jones, O., *The early days of Sirhowy and Tredegar* (Risca, 1969).

Jones, T., *A history of the county of Brecknock* (Brecknock, 1805–1809).

Lancaster, J.Y., and Wattleworth, D.R., *The iron and steel industry of west Cumberland. An historical survey* (Workington, 1977).
Lane, H.C. (ed.), *Derwent Archaeological Society Research Report*, 1 (Matlock, 1973).
Laun, J. van, '17th century ironmaking in south west Herefordshire', *Historical Metallurgy*, 13 (1979), 55–68.
   'Industrial archaeology, 1987. Bringewood furnace and forge site — a reassessment', *Trans. Woolhope Naturalists' Field Club*, 45 (1985–7), 787.
   The Clydach Gorge: industrial archaeology trails in a north Gwent valley (Brecon Beacons National Park Committee, 2nd ed. 1989).
Lead, P., 'The north Staffordshire iron industry 1600–1800', *Historical Metallurgy*, 11 (1977), 1–14.
Lewis, J.H., 'The charcoal-fired blast furnaces of Scotland: a review', *Proc. Soc. Antiquaries of Scotland*, 114 (1984), 433–79.
Lewis, R.A., 'Two partnerships of the Knights: a study of the Midland iron industry in the eighteenth century' (Unpublished University of Birmingham M.A. thesis, 1949).
Lindsay, J.M., 'Charcoal iron smelting and its fuel supply; the example of Lorn furnace, Argyllshire, 1753–1876', *Journal of Historical Geography*, 1 (1975), 283–98.
   'The iron industry in the Highlands: charcoal blast furnaces', *Scottish Historical Review*, 56 (1977), 49–63.
Linsley, S.M., and Hetherington, R., 'A seventeenth century blast furnace at Allensford, Northumberland', *Historical Metallurgy*, 12 (1978), 1–11.
Llewellin, W., 'Sussex ironmasters in Glamorganshire', *Archaeologia Cambrensis*, 3rd series, 9 (1863), 81–119. [1863a]
   'Some account of the iron and wireworks of Tintern', *Archaeologia Cambrensis*, 3rd series, 9 (1863), 291–318. [1863b]
[Llewelyn, T. D.], 'Aberdare in 1853. A translation (with introduction and commentary) by D.L. Davies of the first essay on the history of the parish of Aberdare by Thomas Dafydd Llewelyn', *Old Aberdare*, 2 (1982), 7–85.
Lloyd, H., *The Quaker Lloyds in the industrial revolution* (1975).
Lloyd, J., *The early history of the old south Wales ironworks, 1760–1830* (1906).
Lloyd, J.E. (ed.), *A history of Carmarthenshire*, II (Cardiff, 1939).

Macadam, W.I., 'Notes on the ancient iron industry of Scotland', *Proc. Soc. Antiquaries of Scotland*, new series, 9 (1887), 89–131.
Magilton, J., *The archaeology of Chichester and district 1989* (Chichester, 1990).
Marshall, J.D., and Davies-Shiel, M., *The industrial archaeology of the Lake Counties* (Newton Abbot, 1969).

Martin, E., *Bedlington iron and engine works (1736–1867). A new history* (Newcastle-upon-Tyne, 1974).

Melling, E. (ed.), *Kentish sources: III. Aspects of agriculture and industry* (Maidstone, 1961).

Minchinton, W.E., 'The place of Brecknock in the industrialization of south Wales', *Brycheiniog*, 7 (1961), 1–46.

Mineral Statistics of the United Kingdom of Great Britain and Ireland (Annual, 1852–1913; Geological Survey to 1883, thereafter Parliamentary Papers).

Morris, D., *Hanes Tredegar o ddechreuad y gwaith haiarn hyd yr amser presenol* (Tredegar, 1868).

Morris, W.H., 'Kidwelly tinplate works: 18th century leases', *Carmarthen Antiquary*, 5 (1964–9), 21–4.

Morton, G.R., 'The furnace at Duddon Bridge', *Journal of the Iron and Steel Institute*, 200 (1962), 444–52.

'The products of Nibthwaite ironworks', *The Metallurgist*, 2 (1963), 259–68.

'The reconstruction of an industry', *Journal of the Lichfield & South Staffs. Arch. & Hist. Soc.*, 6 (1964–5), 21–38.

'The technical development of the south Staffordshire iron industry', *Iron and Steel*, 38 (1965), 314–20.

'The use of peat in the extraction of iron from its ores', *Iron and Steel*, 38 (1965), 421–4.

'Some details of an early blast furnace', *Iron & Steel*, 39 (1966), 563–6.

See also Hallett, M.M.

Mott, R.A., 'Early ironmaking in the Sheffield area', *Trans. Newcomen Soc.*, 27 (1949–51), 225–35.

'The early history of Wortley Forges', *Bull. Hist. Metallurgy Group*, 5 (1971), 63–70.

Mutton, N., 'Charlcotte furnace', *Trans. Shropshire Arch. Soc.*, 58 (1965–8), 84–8. [1965–8a]

'The forges at Eardington and Hampton Loade', *Trans. Shropshire Arch. Soc.*, 58 (1965–8), 235–43. [1965–8b]

'Charlcot furnace 1733–1779', *Bull. Hist. Metallurgy Group*, No 6 (1966), 18–48.

'Sites of charcoal blast furnaces at Shifnal and Kemberton, Shropshire, 1972', *Bull. Hist. Metallurgy Group*, 7 (1973), 26–7.

Nixon, F., *The industrial archaeology of Derbyshire* (Newton Abbot, 1969).

Nichols, R. (ed.), *Monmouthshire Medley*, 3 (Pontypool, 1978).

Paar, H.W., 'The furnaces at Coed Ithel and Trellech', *Bull. Hist. Metallurgy Group*, 7 (1973), 36–9.

Paar, H.W., and Tucker, D.G., 'The old wireworks and ironworks of the Angidy valley at Tintern, Gwent', *Historical Metallurgy*, 9 (1975), 1–14.

BIBLIOGRAPHY

Page, R., 'Richard and Edward Knight: ironmasters of Bringewood and Wolverley', *Trans. Woolhope Naturalists' Field Club*, 43 (1982), 7–17.

Palmer, A.N., *John Wilkinson and the old Bersham ironworks* (Wrexham, 1899).

Pape, T., 'How the iron age ended at Leighton Beck and Halton', *Lancaster Guardian and Observer*, 25 Jan. 1959.

Phillips, C.B., 'The Cumbrian iron industry in the seventeenth century', in Chaloner, W.H., and Ratcliffe, B.M. (ed.), *Trade and transport. Essays in economic history in honour of T.S. Willan* (Manchester, 1977).

Phillips, D.R., *The history of the vale of Neath* (Swansea, 1925).

Phillips, M., 'The early development of the iron and tinplate industries of the Port Talbot district', *Trans. Aberafan & Margam District Hist. Soc.*, 5 (1932–3), 11–30.

Pickin, J., 'Excavations at Abbey Tintern furnace', *Historical Metallurgy*, 16 (1982), 1–21. [1982a]

'The ironworks at Tintern and Sirhowy', *Gwent Local History*, 52 (1982), 3–9. [1982b]

Ponsford, M., 'Post-medieval Britain and Ireland in 1990', *Post-Medieval Archaeology*, 25 (1991), 115–70.

Powell, E., *History of Tredegar* (Cardiff, 1885).

Pugh, R.H., *Glimpses of west Gwent. An historical survey of Abercarn and district* (Newport, Mon., 1934).

Raistrick, A., 'The south Yorkshire iron industry, 1698–1756', *Trans. Newcomen Soc.*, 19 (1938), 51–86.

Quakers in science and industry. Being an account of the Quaker contribution to science and industry during the 17th and 18th centuries (2nd ed., Newton Abbot, 1968).

Dynasty of ironfounders. The Darbys and Coalbrookdale (2nd ed., Newton Abbot, 1970).

'The Old Furnace at Coalbrookdale', *Industrial Arch. Review*, 4 (1979–80), 117–34.

Raistrick, A., and Allen, E., 'The south Yorkshire ironmasters (1690–1750)', *Economic History Review*, 9 (1939), 168–85.

Rees, D.M., 'Industrial archaeology in Cardiganshire', *Ceredigion*, 5 (1964–7), 109–24.

Mines, mills and furnaces: an introduction to the industrial archaeology in Wales (National Museum of Wales, 1969).

'Iron', in Jenkins, E. (ed.), *Neath and district. A symposium* (Neath, 1974), 149–65.

The industrial archaeology of Wales (Newton Abbot, 1975).

Rees, W., *Industry before the industrial revolution incorporating a study of the chartered companies of the Society of Mines Royal and of Mineral and Battery Works* (Cardiff, 1968).

Rendell, B., 'A preliminary report on the excavations and research of the Lydney blast furnace complex, 1604–1810', *New Regard of the Forest of Dean*, 2 (1986), 40–45.
Reynolds, P., see Hughes, S.
Rhys, J., 'Hanes hen ffwrnes wynt Caerffili, yn swydd Forgannwg', *Seren Gomer*, 41 (1858), 204–209.
Richards, H.P., 'Caerphilly Furnace: Ffwrnes Caerffili', *Caerphilly*, 1 (1968), 36–45.
Riden, P.J., 'Excavations at Wingerworth ironworks', *Bull. Hist. Metallurgy Group*, 7 (1973), 48.
'The output of the British iron industry before 1870', *Economic History Review*, 2nd series, 30 (1977), 442–59.
'Eighteenth-century blast furnaces: a new checklist', *Historical Metallurgy*, 12 (1978), 36-9.
'Joseph Butler, coal and iron master, 1763–1837', *Derbys. Arch. Journal*, 104 (1984), 87–95.
(ed.), *George Sitwell's Letterbook, 1662–66* (Derbyshire Record Soc., 10, 1985).
'The ironworks at Alderwasley and Morley Park', *Derbys. Arch. Journal*, 108 (1988), 77–107.
'The ironworks at Alderwasley and Morley Park: a postscript', *Derbys. Arch. Journal*, 109 (1989), 175–9.
'The charcoal iron industry in the East Midlands, 1580–1780', *Derbys. Arch. Journal*, 111 (1991), 64–84.
John Bedford and the ironworks at Cefn Cribwr (Cardiff, 1992). [1992a]
'Early ironworks in the lower Taff valley', *Morgannwg*, 36 (1992), 69–83. [1992b]
'Some unsuccessful blast furnaces of the early coke era', *Historical Metallurgy*, 26 (1992) (forthcoming). [1992c]
See also Glover, C.
Roberts, D.H.V., 'Another early iron furnace at Ponthenri', *Y Gwendraeth*, 2 (1979–80), 16.
Rowley, R.T., 'Bouldon Mill: 700 years of rural industry', *Shropshire Magazine* (February 1966), 28–9.

Sale, R. McN., 'Excavations at Bersham ironworks, 1976', *Trans. Denbighs. Hist. Soc.*, 27 (1978), 150–77.
Saville, R.V.: *see* Crossley, D.W.
Scrivenor, H., *History of the iron trade, from the earliest records to the present period* (1841).
Schubert, H.R., 'The old blast furnace at Maryport, Cumberland', *Journal of the Iron & Steel Institute*, 172 (1952), 162.
'The King's Ironworks in the Forest of Dean, 1612–74', *Journal of the Iron & Steel Institute*, 173 (1953), 153–62.
History of the British iron and steel industry from c. 450 B.C. to A.D. 1775 (1957).

Smith, D.M., *The industrial archaeology of the East Midlands (Nottinghamshire, Leicestershire and the adjoining parts of Derbyshire* (Dawlish and London, 1965).

Spavold, J. (ed.), *At the sign of the Bulls Head. A history of Hartshorne and its enclosure* (S. Derbys. Local History Research Group, 1984).

Standing, I.J., 'Blakeney charcoal blast furnace site', *New Regard of the Forest of Dean*, 2 (1986), 60–62.

Standing, I.J., and Coates, S.D., 'Historical sites of industrial importance on Forestry Commission land in Dean', *Gloucs. Industrial Arch. Soc. Journal* (1979), 16–20.

Stell, G.P., and Hay, G.D., *Bonawe Iron Furnace* (Edinburgh, 1984).

See also Hay, G.D.

Straker, E., *Wealden iron* (1931).

Styles, R., 'Elmbridge Furnace, Oxenhall', *Gloucestershire Historical Studies*, 5 (Bristol University, 1972), 2–11.

Symons, M.V., *Coal mining in the Llanelli area. Volume one: 16th century to 1829* (Llanelli Borough Council, 1979).

Taylor, E., 'The seventeenth-century iron forge at Carey Mill', *Trans. Woolhope Naturalists' Field Club*, 45 (1985–7), 450–68.

Thomas, B., 'Iron-making in Dolgellau', *Journal of the Merioneth Hist. & Record Society*, 9 (1981–4), 474–5.

Thomas, C., 'Industrial development to 1918', in Merthyr Teachers' Centre Group, *Merthyr Tydfil. A valley community* (Cowbridge, 1981), 271–331.

Thomas, D., *Hanes Pontiets a'r cylch* (Llanelli, 1905).

Tomlinson, H., 'Wealden gunfounding: an analysis of its demise in the eighteenth century', *Economic History Review*, 2nd series, 29 (1976), 383–400.

Treadwell, J.M., 'William Wood and the Company of Ironmasters of Great Britain', *Business History*, 16 (1974), 97–112.

Trinder, B., *The industrial revolution in Shropshire* (Chichester, 2nd ed. 1981).

Truran, W., *The iron manufacture of Great Britain theoretically and practically considered* (1855).

Tucker, G., and Wakelin, P., 'Metallurgy in the Wye valley and South Wales in the late 18th century. New information about Redbrook, Tintern, Pontypool and Melingriffith', *Historical Metallurgy*, 15 (1981), 94–100.

See also Paar, H.W.

Turbutt, G., *A history of Shirland and Higham, Derbyshire* (Shirland, 1977).

Tylecote, R.F., 'Blast furnace at Coed Ithel, Llandogo, Mon.', *Journal of the Iron & Steel Institute*, 204 (1966), 314–9.

'A survey of iron and steel making sites in the Tyne–Wear area of the United Kingdom', *Canadian Mining & Metallurgical Bulletin*, 76 (1983), 90–101.

Tylecote, R.F., Banks, A.P., Wattleworth, D.R., and Ashmore, B.G., 'Maryport blast furnace: post mortem and reconstruction', *Journal of the Iron & Steel Institute*, 203 (1965), 867–74.

Wakelin, P., *see* Tucker, D.G.

Wallis, J., *The natural history and antiquities of Northumberland and so much of the county of Durham, as lies between the rivers Tyne and Tweed* (1769).

Wanklyn, M.D.G., 'Iron and steelworks in Coalbrookdale in 1645', *Shropshire Newsletter*, 44 (1973), 3–6.

'Industrial development in the Ironbridge Gorge before Abraham Darby', *West Midlands Studies*, 15 (1982), 3–7.

Wattleworth, D.R., *see* Lancaster, J.Y.; Tylecote, R.F.

Whittick, C., 'Wealden iron in California', *Wealden Iron*, 2nd series, 12 (1992), 29–62.

Wight, M., *see* Cantrill, T.C.

Willan, T.S., *The Muscovy merchants of 1555* (Manchester, 1953).

Williams, D.T., *The economic development of the Swansea district to 1921* (Cardiff, 1940).

Williams, L.J, 'A Welsh ironworks at the close of the seventeenth century', *National Library of Wales Journal*, 11 (1959–60), 266–84.

'The Welsh tinplate trade in the mid-eighteenth century', *Economic History Review*, 2nd series, 13 (1960–1), 440–9.

'A Carmarthenshire ironmaster and the Seven Years War', *Business History*, 2 (1960–1), 32–43.

Williams, M.I., 'The port of Aberdyfi in the 18th century', *National Library of Wales Journal*, 18 (1973–4), 95–134.

Wilson, A., 'The excavation of Clydach Ironworks', *Industrial Archaeology Review*, 10 (1988), 16–36.

Index

Abbey Tintern furnace: *see* Tintern
Abdon furnace, Shropshire 55, 56
Aberavon forge, Glam. 19, 25
Abercarn furnace, Mon. 11–12
Aberdyfi: *see* Dyfi
Aberfoyle furnace, Perthshire 145
'Abergwythen': *see* Abercarn
Abernethy furnace, Inverness-shire 145
Achray: *see* Aberfoyle
Acton, Clement 78
Ainslie: *see* Harrison Ainslie & Co.
Alderwasley furnace, Derbys. 87–8
Allensford furnace, Northumberland 123–4
Allsop, Russell 100
Alvington forge, Gloucs. 47
Argyll, Duke of: *see* Campbell, John
Argyll furnace: *see* Craleckan
Armstrong, Charles 150
Arundel, Earl of: *see* Howard, Thomas
Ash, Martin 95
Ashburnham, John
   John, 2nd Earl of Ashburnham etc 131
   John, 3rd Baron and 1st Earl Ashburnham 133
Ashburnham furnace, Sussex 131–2, 134
Ashton-in-Makerfield, Lancs. 108
Astley furnace, Worcs. 75
Aston furnace, Warwicks. 75-6
Atkinson, — 121
   J. 117
   see also Hartley, Atkinson & Co.
Attwood, — 78
Aubrey, John 133
   Thomas 33
Avenant, Richard 46, 47-8
Ayrey, John 112

Backbarrow Co. 114, 116, 120
Backbarrow furnace, Lancs. 107–8, 111
Backhouse, James 119
Bacon, Anthony 12, 19, 21
Bagshaw, Samuel 81
Baker, John 136
   Robert 136
Baldwin, Richard 63

Banckes, John 100
Bank furnaces, Yorks. 99–100, 102
Barden furnace, Kent 132
Barepot: *see* Seaton
Barlow, Francis 102
Barlow furnace, Derbys. 88
Barnby furnace, Yorks. 100–101
Baskerville, Thomas 43
Bathurst, Benjamin 44
   family, Earls Bathurst 95
Baugh, Edward 63
Beaufort, Duke of: *see* Somerset, Henry
Beckett, John 151
Beckley furnace, Sussex 132
Bedburn, Co. Durham 128
Bedford, John 12
Bedgebury furnace, Kent 132
Bedlington furnace, Durham/Northumberland 124–5
Bedwellty furnace, Mon. 13
Beech furnace, Sussex 133
Bell, — 71
Bellingham furnace, Northumberland 125–6
Benge, William 138
Bersham furnace, Denbighs. 65–8, 117
Bertram family 123
Bigsweir furnace, Gloucs. 35
Birkett, Miles 112
Birtley, Co. Durham 128
Bishop Aucklamd, Co. Durham 128
Bishopswood furnaces, Gloucs./Herefs. 36-7
Blackpool furnace and forge, Pembs. 13, 17
Blakeney furnace, Gloucs. 37
Blount, Edward 55
   Henry 55
   Sir Walter 55
Bodnant: *see* Conwy
Boevey, Mrs Catherine 39
   see also Crawley-Boevey
Bonawe furnace, Argyllshire 146-8
Botfield, Thomas 60
Bouldon furnace, Shropshire 55-6
Bourne, — 135
   James 79, 141
Bowen, William 134

165

INDEX

Boycott, Francis 62
   William 62
Bradshaw, — 112
Braine, Capt. 47
Brecon furnace, Brecs. 13–15, 28
Brede furnace, Sussex 133
Bretland, Thomas 97
Bretton furnace, Yorks. 101
Bridge, Edward 68, 69
   William 68, 69
Briggs family 55
Bringewood furnace, Shropshire/Herefs. 56
Britton, Col. 101
Brockweir, Gloucs. 35
Bromford forge, Warwicks. 75
Brooke, Sir Basil, sen. and jun. 59
Brown, Copley 49
   Jeremiah Sharp 49
Browne, George 132, 136, 138
   John 136, 137
   family 136, 137
Brudenell, Ethelreda 59
   Sir Edmund 59
Bryn Coch furnace, Glam. 15, 24
Buckle, George 146
Bullock, William 93
   family 93
Bulwell forge, Notts. 90
Burdett, Elizabeth 89
'Burhamfold': *see* Fernhurst, Burningfold
Burley, Nicholas 113
Burningfold furnace, Surrey 133
Burton-on-Trent forge, Staffs. 91
Bute, Mary Countess of: *see* Montague, Mary Wortley
Butler, John 135
   Joseph, sen. and jun. 97

Cadell & Eddington 150
Caerphilly furnace, Glam. 15–17, 22, 32
Calke, Derbys. 93
Camden, Earl: *see* Pratt, John Jeffreys
Campbell, John, Duke of Argyll etc 146
Canaston Bridge: *see* Blackpool
Cannock Chase furnaces, Staffs. 76–7
Cannop furnace, Gloucs. 37
Carburton forge, Notts. 81, 94
Cardiff, Glam., New Forge at 21
Cardiff, Lord: *see* Stuart, John, Marquess of Bute etc
Carey Mill forge, Herefs. 53, 78
Carmarthen furnace, Carms. 17–19, 29
Carr Mill forge, Lancs. 108

Carron Co. 150
Catchmay family 35, 39
Cathcart family, Barons Cathcart 152
Cavendish, William, 1st Duke of Newcastle etc 90
   William, 5th Duke of Devonshire etc 94
Chadwick, John 69
Challenor, John 13
   William 13
Champion, Nehemiah 151
Chapel furnace, Yorks. 102
Charlcotte furnace, Shropshire 57–8
Chepstow, Mon. 151
Chester-le-Street furnace, Co. Durham 126–8
Chetle, Peter 29
   Thomas 29
Chetwynd, Richard 83
   Walter 82, 83
   William 83
Chetwynd-Talbot, Charles Chetwynd, 3rd Earl Talbot of Hensol etc 24, 27
Childe family 57
Christian, — 121
Church Preen, Shropshire, proposed furnace at 59
Churchill, John 79, 141
Clay, John 102
Clayton, Robert 43
   William 94
Cleator furnace, Cumb. 109
Clee Hill furnace, Shropshire 60
Cleobury Mortimer forge, Shropshire 60
Cliviger Ponds, Lancs. 113–14
Clutton, Ralph 136
   William 136
Clutton & Co. 136
Clutton & Durrant 139
Clydach Ironworks, Brecs. 23–4
Coalbrookdale Co. 63
Coalbrookdale furnace, Shropshire 58, 59–60
'Coalford' furnace, Shropshire 55–6
Cockshutt, John 100, 102
   see also Cook & Cockshutt
Coed Ithel furnace, Mon. 38–9, 49, 52
Cole, Thomas Butler 113
Coles, William 25
Coles, Lewis & Co. 15, 22, 25
Collins, Thomas 141
   William 138
Colnbridge forge, Yorks. 113
Colnbrook Co. 101
Company of Mineral & Battery Works 25
Conwy furnace, Denbighs. 68

INDEX

Cook & Cockshutt, Messrs 101
Cookburn, — 99, 100
Cookson, Isaac 126
   John 127
   Joseph 126
   N.C. 126
   family 114
Copley, Lionel 93, 102, 103, 105
Corfield, Thomas 59
   William 59
Cornbrook furnace, Shropshire 60
Cotton, — 83, 84
   Daniel 82, 85, 109
   Thomas 76, 101
   William 61, 72, 82, 99, 100–101
   William Westby 61, 101, 103
Courthope, Alexander 136, 138
   family 136
Coushopley furnace, Sussex 134
Cowden furnace, Kent 134, 141
Cox, Joseph 78
Cradley furnace, Worcs. 77
Craleckan furnace, Argyllshire 148–50
Cramond ?furnace, Midlothian 150
Craven, William, 1st Baron and 1st Viscount Craven, 1st Earl of Craven 56
   William, 7th Baron Craven of Hampsted Marshall etc 60
Crawley, T.B.: *see* Crawley-Boevey, Thomas
Crawley-Boevey, Thomas 39-40
   see also Boevey
Crawshay, Richard 12
   William 21
Crewe, John 83
Crich Chase forge, Derbys. 94
Croft, Richard 77
   William 77
Croker, Philip 17
Crossfield, Stephen 107
   William 116
Crowhurst furnace, Sussex 134
Crowley, Ambrose 33
   John 134
   Richard 33
   family 124, 131
Cuckney forge, Notts. 95
Culnakyle: *see* Abernethy
Cunsey Co. 109–110
Cunsey furnace, Lancs. 109
Cwm Ffrwd-oer: *see* Pontypool
Cwmddyche: *see* Llanelli
Cwmdwyfran forge, Carms. 17
Cwrtrhydhir: *see* Longford Court

Daniel, Thomas, jun. 44
   Thomas sen. 14, 44
Darby, Abraham, jun. 59
   sen. 59, 68, 85
   family 62
Darbyshire, John 12
Darwell furnace, Sussex 134–5
Davidson, John 151
Davies, James 47
Davies & Co. 47
Davison, — 123
Dearman, J. Petty 121
   Richard 121
Delves, Sir Thomas 81
Derby, Earl of: *see* Smith-Stanley, Edward
'Derrington': *see* Doddington
Derwentcote, Co. Durham 123
Devereux, Robert, 2nd Earl of Essex etc 43
Devonshire, Duke of: *see* Cavendish, William
Dickin, Thomas 100–101, 113
Dickins, John 59
Disley furnace, Cheshire 81
Dixon, John 142, 146
Doddington furnace, Cheshire 81
Dol-gûn furnace, Merionethshire 68–9
Dolgellau: *see* Dol-gûn
Dolobran forge, Mont. 65, 68
Dowlais furnace, Glam. 21–2
Downing, William 56
   Zachary 29
Downing & Cooley 56
Drinkal, George 116
Duddon Co. 148
Duddon furnace, Cumb. 109–11
Dudley, Lord: *see* Sutton (or Dudley), Edward
Dunsfold: *see* Burningfold
Durrant, Samuel 136
   see also Clutton
Dyfi furnace, Cards. 69–71
Dyke, William 137

Eade & Wilton 136
Eckington, Derbys. 89
Eddington: *see* Cadell & Eddington
Edwards, Lewis 69
Effingham, Earl of: *see* Howard, Richard
Eglwysbach: *see* Conwy
Eglwysfach: *see* Dyfi
Elmbridge furnace, Gloucs.: *see* Newent
Erbury, Thomas 31
Essex, Earl of: *see* Devereux, Robert
Ewhurst furnace, Sussex 135, 139

INDEX

Falkner, Warine 76, 83
Fallowfield, William 82
Farnden, Peter 132, 133, 134, 141
Fell, John 102
Fenwick, Nicholas 123–4
Fernhurst furnace, Sussex 135
Finch, Francis 45
   Heneage, 3rd Earl of Winchilsea etc 134
Flaxley furnace, Gloucs. 39-40
Fletcher, William 84
Fold, — 142
Foley, Paul 36, 37, 42, 46, 47
   Philip 57, 63
   Richard 75, 77, 78, 83
      of Longton 83
      of Stourbridge 83
   Thomas 40, 43, 45, 48, 63, 77
      5th Baron Foley of Kidderminster 36
   family 36, 37, 39, 42, 45–8, 62, 78, 81, 83–5, 89, 132, 136
Ford, Richard 62, 63, 65, 119, 146
   William 119, 142, 146
Foremark furnace, Derbys. 89
Fownes, William 61, 62, 100
Fox, Shadrach 59
Foxbrooke furnace, Derbys. 89, 94
Frecheville family 93–4
Freeth, Sampson 121
   Samuel 121
Frith furnace, Sussex 135
Frizington furnace, Cumb. 112
Fuller, John 137, 142, 143
   family 137, 142, 143
Furnace, Argyllshire: *see* Cralecken
Furnace, Cards.: *see* Dyfi
Furnace, Carms.: *see* Llanelli
Furnace End: *see* Pool Bank furnace, Warwicks.
Furnace Grange: *see* Grange furnace, Staffs.

Gage, Henry, 3rd Viscount Gage of Castle Island etc 46–7
   William Hall, 2nd Viscount Gage of Castle Island etc 46
Gairloch, Ross & Cromarty, furnaces at 150
Galbraith, Arthur 150
Gale, John 116
   Leonard, sen. and jun. 141
Gateshead, Co. Durham, foundry at 127
Gee, Joshua 65
George & Co. 60
Gerard, William 108
Getley: *see* Reynolds, Getley & Co.

Gibbons, Benjamin 77
   Thomas 77
   William 77
Gibson, Edward 116
Giles, Benjamin 56
Glanfraid, Cards., forge at 69
Glascoed, Mon. 26
Glen Kinglass furnace, Argyllshire 150–1
Gloucester Bank 47
Gloucester furnace: *see* Lamberhurst
Glover, Joshua 12
   Samuel 12, 19
Goatfield: *see* Cralecken
Goldney, Thomas 62, 63, 65
Gonning, John 47
Goodyer, James 135
Gordon of Kirkconnel, John 145
Gosling, John 108
Gott, Miss 132
   Samuel 132, 133, 134, 138
Graham of Kilmannan, James 145
Graham's Miln: *see* Aberfoyle
Grange furnace, Staffs. 78
Gravetye furnace, Sussex 135–6, 139, 143
Greenhough, John 32
Griffiths, John 12
Grove, James 57
Grundy, Hugh 29
   Lucy 29
Gunne, William 40
Guns Mill furnace, Gloucs. 40–2
Guy, David 141
Gyles, Revd John 134

Hackett, Thomas 32
Hales furnace, Halesowen, Shropshire/Worcs. 78–9
Hall, — 142
   Benedict 46
   Benjamin 12
   Daniel 82
   Edward 76–7, 83, 85, 101, 108, 109
   Thomas 82
   William 47
Hall & Co. 81, 83
Hall, Kendall & Co. 110
Halton Co. 114
Halton furnace, Lancs. 112–13
Hampton Loade furnace, Shropshire 61
Hamsell furnace, Sussex 136
Hanbury, Capel 12, 23
   John 23, 24, 30, 33, 49
   Richard 11, 25, 30

INDEX

family 30, 131
Harford, Daniel (?) 44
  James 17
  John Scandrett 44
  Truman 17
Harford, Partridge & Co. 17, 27, 46–7
Harley forge, Shropshire 61, 62
Harrison, — 132, 143
  William 133, 136, 138
  family 137, 138
Harrison Ainslie & Co. 109–10, 119, 143, 146
Harrison, Jukes & Co. 140
Hart, Richard 32
Hartleton, Herefs.: see Linton
Hartley, Thomas 116
Hartley, Atkinson & Co. 117
Hartshorne furnace, Derbys. 89–90
Harvey, Benjamin 65
  Charles 133
  Thomas 72
Hastings, Francis, 10th Earl of Huntingdon etc 91
Hatton & Co. 112, 114
Hawes, William 133
Hawkhurst furnace, Kent 136, 137
Hawkins, Ann 65
  John 65
Hawks, Longridge & Co. 124–5
Hay, Richard 133
Heathfield furnace, Sussex 137, 142
Heighley furnace, Staffs. 82
Herbert, William, 3rd Earl of Pembroke etc 37, 46, 48
Heyford, Dennis 102, 103, 123
Hicks, Spedding & Co. 121
Higgons, William 73
Hill, Caleb 83
  William 83
Hints forge, Staffs. 79
Hirwaun furnace, Brecs. 12, 14, 19–22, 28
Hodgson, John 128
Holbeck, Richard 61
Holme Chapel furnace, Lancs. 113–14
Holte, Sir Thomas 75
Homfray, Francis 44
  Jeremiah 44
Hooper, Charles 140
Horsmonden furnace, Kent 137
Horsted Keynes furnace, Sussex 137
Horton furnace: see Leek
Howard, Richard, 4th Earl of Effingham etc 103
  Thomas, 21st or 14th Earl of Arundel 102
Howbrooke: see Lydbrook
Hughes, Lewis 29
Hunloke, James 97
  family 95–7
Huntingdon, Earl of: see Hastings, Francis
Hunwick, Co. Durham 128
Hurt, Francis 87
Hussey, Thomas 138

Ifton Rhyn furnace, Shropshire 72
Imbhams furnace, Surrey 138
Invergarry furnace, Inverness-shire 151
Iremonger, Gregory 32
Iridge furnace, Sussex 138
Irvine, John 145
Ivie, Daniel 72

James, Eleanor 47
Jars, Gabriel 114, 117, 120
Jeffreys, Richard 14
Jenkins' furnace: see Pontyrhun
Jennens, Charles 89
  Humphrey 75, 79, 90
  John 75, 88, 90, 94, 97
  sen. and jun. 79
Jewkes: see Harrison, Jukes & Co.
Jodrell family 81
Johnson, Samuel 108
  see also R. Johnson & Newphew Ltd
Jones, Hugh 15
  Richard 32
  William 61
Jones & Rowland 73
Jordan, — 78
  Edward 49
Jukes, George 140–1
  William 140–1
  see also Harrison, Jukes & Co.

Kelsall, John 68
Kemberton furnace, Shropshire 61
Kemeys, Sir Charles 32
Kendall, — 85, 109
  Edward 61, 76, 77, 79
  Jonathan, & Co. 148
  family 68, 69, 81
Kendall, Latham & Co. 110
Kendall & Co. 71, 81
  see also Hall, Kendall & Co.
Kenley furnace, Shropshire 61, 62
Kettlewell, William 150–1
Kidwelly forge, Carms. 17

INDEX

furnace: *see* Ponthenri
Kimberworth furnace, Yorks. 102
Kingston, Earl of: *see* Pierrepont, Robert
Kirkby-in-Ashfield furnace, Notts. 90
Kirkstall forge, Yorks. 100–101, 113
Knight, Edward 55, 57, 75, 78
    Francis 63
    Ralph 55, 57
    Richard 39, 55, 56, 57, 63, 72, 78
        jun. 63
    Richard Payne 56
    Robert 139
Knight & Son 78
    see also Stour Partnership
Knott, George 119, 120, 146
    L.G. 146
    Michael 119
Knottingley forge, Yorks. 99
Kyrle, Sir John 53

Lacon, Sir Francis 57
    Rowland 61 57
    family 62
Lamberhurst furnace, Kent 138
Latham, Joseph 110
    Richard 110
    William 110
    see also Kendall, Latham & Co.
Lawton furnace, Cheshire 82, 83, 84
Laytons, Mon. 48–9
'Lea' furnace: *see* Disley
Leek furnace, Staffs. 82
Leeke, Sir Francis 91
Legas, — 143
    John 138
Leicester, Earl of: *see* Sidney, Philip
Leighton furnace, Lancs. 112, 114, 115
Leighton furnace, Shropshire 62, 63
Lettsom, John Coakley 24
Leveson-Gower family 83
Lewis, Francis 23
    Thomas 21-2, 25, 26
    William 27
    Wyndham 27
    see also Coles, Lewis & Co.
Lewthwaite, William 116
Ley: *see* Tulk Ley & Co.
Lightmaker, Edward 137
Linton furnace, Herefs. 42–3
Little Clifton furnace, Cumb. 114–16, 127
Lizard forge, Shropshire 82
Llancillo forge, Herefs. 23, 47
Llandyfân furnace and forge, Carms. 22, 29

Llanelli furnace, Carms. 22, 23
Llanelly furnace, Brecs. 23–4
'Llanfrede' forge, Cards. 69
Llewellyn, John 33
Lloyd, Charles 65
    Edward 72
    Sir Richard 65
    family 90
Llygad Cynon: *see* Pontbrenllwyd
Lochaber, Inverness-shire, iron made at 151
Longford Court furnace, Glam. 15, 24
Longhope furnace, Gloucs. 43
Longmore, John 56
Longridge & Co. 125
    see also Hawks, Longridge & Co.
Lorn: *see* Bonawe
Low Wood furnace, Lancs. 116
Lowbridge, Thomas 72
Lowe, Humphrey 77, 78
    Joseph 81
Lowther, Sir James 121
Lydbrook furnace and forge, Gloucs. 43, 46
Lydney furnace, Gloucs. 44–5
Lye forge, Worcs. 77
Lyttelton, Sir Thomas 78
    William Henry, 3rd Baron Lyttelton of Frankley, 1st Baron Westcote etc 78

Macdonald of Invergarry, John 151
Machell, — 120
    James 114
    John 107, 116
Machen (Rhydygwern) forge, Glam. 15, 19, 32
Machynlleth forge, Cards. 71
Madeley furnace, Staffs. 82, 83, 84
Maling, William 124
Mander, John 75, 97
Manning, Charles 140
Mansel, Sir Thomas, 1st Lord Margam 25
Maresfield, Sussex 139
Margam, Lord: *see* Mansel, Sir Thomas
Marshalls furnace: *see* Old Forge
Maryport furnace, Cumb. 116–17
Masbrough furnace, Yorks. 103
Masters & Raby 143
Mather, Walter 90, 94, 97
    family 90
Matthews, George 97, 138
    William 49
Maybery, — 28
    Charles 28
    John 14, 17, 19, 21

INDEX

Mary 19
Thomas 14
Mayfield furnace, Sussex 139
Meir Heath furnace, Staffs. 82, 83, 83–4
Melbourne furnace, Derbys. 90–1, 93
Melincourt furnace, Glam. 15, 24–5
Melingriffith tinplate works, Glam. 27
Miers, John 24, 25, 33
Mill Place furnace, Sussex 139
Milner, Samuel 68
Mitton forge, Worcs. 113
Monkswood furnace, Mon. 25–6
Montague, Mary Wortley, Countess of Bute and Baroness Mount Stuart 99–100
Moore, Hugh 65
  Richard 65
  Thomas 13
Morgan, Anthony 29
  Christopher 17
  Elizabeth 29
  John of Carmarthen 19, 29
    of Tredegar 15, 17
  Lucy (née Grundy) 29
  Robert 17, 19, 29
  Thomas of Carmarthen 29
    of Ruperra 15, 22
    of Tredegar 32
  Sir William 15
Morley Park furnaces, Derbys. 88
Mount Stuart, Baroness: *see* Montague, Mary Wortley
Murphey, Roger 150
Mushet, David 49
Myddleton, — 66
  Richard 73
  Sir Thomas 72
  family 72
Mynd, John 36
Mynors, Robert 47
  family 47
Mytton, Thomas 57

Neath Abbey Ironworks, Glam. 15, 33
Nechells Park slitting-mill, Warwicks. 75
Netherhall: *see* Maryport
Nevile, Francis 102
  Gervase 93
New Forge: *see* Cardiff
New Willey Co. 63
Newborough, Joshua 62
Newcastle, Duke of: *see* Cavendish, William
Newcastle-upon-Tyne, foundry at 127
Newent furnace, Gloucs. 41, 45–6

Newland Co. 116, 119, 146
Newland furnace, Lancs. 118–19
Newport, Francis, 2nd Baron Newport of High Ercall etc 62
  Sir Richard, 1st Baron Newport of High Ercall 62
Nibthwaite furnace, Lancs. 115, 119–20
Nicoll, John 134
North Park: *see* Fernhurst
North Wingfield furnace, Derbys. 91–2
Northiam furnace, Sussex 139
Norton furnace and forge, Derbys. 93

Old Forge furnace, Sussex 139
Old Furnace, Herefs.: *see* St Weonards
Oliphant, John 107
Oulton furnace: *see* Vale Royal
Over Whitacre, Warwicks. 79
Oxenhall, Gloucs.: *see* Newent

Paget, Thomas, 3rd Baron Paget 76
Pallingham furnace, Sussex 140
Park Hall: *see* North Wingfield
Parkend furnace, Gloucs. 46
Parsons, Richard 33
Parsons & Co. 33
Partridge, John 36
  sen. and jun. 46
  William 36
  see also Harford, Partridge & Co.
Patrickson, Richard 109
Payne family 63
Payton, Henry 68
  John 68
Pelham family 143
Pemberton, Thomas 97
Pembroke, Earl of: *see* Herbert, William
Penderyn: *see* Hirwaun, Pontbrenllwyd
Pendrill, — 36, 45
Penkherst family 133
Penny, William 120
Penny Bridge furnace, Lancs. 120
Pentyrch furnace, Glam. 21, 26–7
Perry, George 108
Peterchurch forge, Herefs. 47
Pidcock, John, sen. and jun. 44
  Robert 44
  Thomas 44
Pierrepont, Robert, 1st Earl of Kingston etc 95
Pilsley: *see* North Wingfield
Pippingford furnace, Sussex 140

171

INDEX

Pipton forge, Brecs. 14, 28
Plas Madoc furnace, Denbighs. 72
Platt, John 44
Plot, Robert 83
Plymouth Ironworks, Glam. 31–2
Polhill, William 141
Pont-y-blew forge, Denbighs. 65
Pontbrenllwyd furnace, Brecs. 28
Pontgwaithyrhaiarn: see Bedwellty
Ponthenri furnace, Carms. 17, 29–30
Pontrilas forge, Herefs. 47
Pontygwaith forge, Glam. 31–2
Pontymoel: see Pontypool
Pontypool furnace, Mon. 12, 30–1
Pontyrhun furnace, Glam. 31–2
Pool Bank furnace, Warwicks. 79
Popkin, Robert 33
   Thomas 25
Porthcasseg, Mon. 39, 48–9
Portrey, Christopher 33
Postlethwaite, James 116
   Miles 112
   William 116
Pounsley, — 140
Pounsley furnace, Sussex 140
Powell, Gabriel 33
   John 47
Pratt, James 15
   John Jeffreys, 2nd Earl Camden etc 14
   Samuel 15
Preston, Sir Thomas 107
Price, Mary 27
   Nicholas, jun. 27
   Thomas 26
Prickett: see Wright & Prickett
Priory Mill: see Carmarthen
Probert family 51–2
Pytt, Rowland, jun. 46
   sen. 25, 44, 46
   sen. or jun. 49

R. Johnson & Nephew Ltd 87–8
Raby, Alexander 22, 23
   Edward 143
Raby & Co. 136, 143
Raby & Rogers 136, 143
   see also Masters & Raby
Raikes, — 44
Rawlinson, Job 116
   Thomas 151
   William 107
Rea, William 48
Redbrook furnace, Gloucs. 46–7

Renishaw slitting-mill, Derbys. 89, 94
Reynolds, Richard 46
   sen. 14
Reynolds, Getley & Co. 44
Rhydygwern forge, Glam.: see Machen
Richardson, Revd J. 133
Rigg, Thomas 119
Rigley, — 108
Roberts, Edmund 11–12
   John 140, 141
Robertsbridge furnace, Sussex 135, 140–1
   steelworks 32
Robin Hood's Bay, Yorks. 126
Rockley, Sir Francis 102
Rockley furnace, Yorks. 102, 103–5
Rodmore furnace, Gloucs. 47
Rogers: see Raby & Rogers
Rogerstone forge, Mon. 12
Rotherham, Yorks. 103
Routh, Samuel 112
Rowland, John 72
   family 73
   see also Jones & Rowland
Ruabon furnace, Denbighs. 72–3
Rugeley slitting-mill, Staffs. 83
Rushall furnace, Staffs. 79
Ruston, John 44

St Weonards furnace, Herefs. 47–8
Savile, Thomas, 2nd Viscount Savile etc 100
'Sawley': see Sowley
Scarcliffe, Derbys. 95
Scarlett, Benjamin 134
Scarletts furnace, Kent 141
Scudamore, Sir James 42
   Sir John 53, 78
Seacroft furnace, Yorks. 105
Seaton furnace, Cumb. 120–1
Senhouse, Humphrey 116
Senior, William 91, 92
Seys, Richard 33
Shaw, John 61
Sheffield, Yorks. 105
Sheinton forge, Shropshire 62
Shewen, Daniel 22
   Joseph 22
Shirland, Derbys. 94
Shirley, Sir George 93
Shore, Samuel 101
Shrewsbury, Earl of: see Talbot, George; Talbot, Gilbert
Sidney, Sir Henry 32
   Philip, 3rd Earl of Leicester etc 140

172

INDEX

Silvester, John 113
Simpson, John 102
  William 102
Sitwell, George 89, 91, 93, 94
Slaney, Robert 61
Smith, John 83–4, 145
  Sir Richard 63
Smith-Stanley, Edward, 12th Earl of Derby 114
Snepp, John, sen. and jun. 140
Socknersh furnace, Sussex 141
Somerset, Henry, 10th Earl and 1st Marquess of Worcester 48
  Henry, 3rd Marquess of Worcester and 1st Duke of Beaufort 49
  Henry, 5th Duke of Beaufort etc 51
Soresby, William 97
Soudley furnace, Gloucs. 48
Southampton, Earl of: *see* Wriothesley, Henry
Sowley furnace, Hants. 142
Spedding & Co. 121
  see also Hicks, Spedding & Co.
Spencer, John (fl. 1713–27) 113, 102
  John (fl. 1774) 99, 101
  John son of John Spencer of Criggion 100–101
  John son of Randolf 100
  Randolf 100
  Walter 100
  William 94, 101, 102
Spencer family ironworks 89, 90, 94, 95, 99-103, 105
Spooner, Abraham 75–6
Stainborough, Yorks. 103
Staunton Harold, Derbys. 93
Staveley furnace and forge, Derbys. 90, 93–4, 102
Stevens, Alice 32
Stevenson, John 83
Stour Partnership 12, 15, 17, 23, 29, 30, 33, 36, 40, 44, 45, 46, 49, 60, 63, 68, 69, 71, 72, 76, 77, 78, 79, 81, 83, 107, 108, 109, 110, 112, 114, 116, 117, 119, 120, 121, 142, 146, 148
Strangworth forge, Herefs. 56
Stream furnace, Sussex 142
Street furnace, Cheshire 84
Strontian leadworks, Argyllshire 151
Stuart, John, Baron Cardiff of Cardiff Castle, 1st Marquess of Bute etc 21
Sunderland, John 116
  Thomas 116
Sutton (or Dudley), Edward, 5th Baron Dudley 77
Swallow, Richard 102
Swedenborg, Emanuel 138

Talbot, George, 6th Earl of Shrewsbury 53, 88
  Gilbert, 7th Earl of Shrewsbury etc 94
  William, 2nd Lord Talbot of Hensol, 1st Earl Talbot 26
  see also Chetwynd-Talbot, C.C.
Talycafn: *see* Conwy
Tanner, Benjamin 13–14
  David 14, 23, 30, 44, 47, 49, 51
  William 14
Tanner's Forge: *see* Brecon furnace
Tarrioch furnace, Ayrshire 152
Teanford furnace, Staffs. 84
Tern forge, Shropshire 72
Terrioch: *see* Tarrioch
Thomas, Edward 69
Thomlinson, William 124
Thompson, John 61
  Robert 49, 51
  William 49
Tilgate furnace, Sussex 142
Tilsop furnace, Shropshire 62
Tintern furnace, Mon. 39, 48–51
Toadhole furnace, Derbys. 94
Todd, — 142
Tongwynlais furnace, Glam. 32–3
Tredegar Park forge, Mon. 15, 19
Trellech furnace, Mon. 51–2
Trosnant: *see* Pontypool
Trostrey forge, Mon. 12
Tubman, Edward 116
Tulk Ley & Co. 121
Turner, John 82

Upper Norncott, Shropshire 55
Uppington forge, Shropshire 62

Vale Royal furnace, Cheshire 84–5
Vaughton, Christopher 75
  Riland 75, 97
Vauxhall furnace, Kent 142
Venables-Vernon, George, 2nd Lord Vernon etc 24
Vernon, Lord: *see* Venables-Vernon, George, Lord Vernon etc
  Ralph 68, 69

Wadsley furnace, Yorks. 105
Waldron furnace, Sussex 143
Walford: *see* Bishopswood

INDEX

Walker, Francis 56, 62
  Job 56
  Richard 56
  William 62
  family 103
Ward, — 79, 125
Warren furnace, Sussex 143
Warsash furnace, Hants. 143
Weaman, Phelicia 75
Weardale, Co. Durham, furnace in? 128
Weare, John Fisher 44
Webster, Sir Thomas 133, 140
  Sir Whistler 141
Weld, John 62–3
Welford, Richard 127
Wellington, Richard 13–14
Wenham, John 93
Westcote, Lord: *see* Lyttelton, William Henry
Westerne, Maximilian 133
  Thomas 131, 133
Whaley furnace, Derbys. 95
Wheeler, John 46, 47–8, 83, 89
  family 77
Whitchurch furnace, Herefs. 53
White, George, jun. 49
  Richard 49
Whitecroft forge, Gloucs. 46
Whitehill: *see* Chester-le-Street
Whitland furnace and forge, Carms. 17, 29
Whitmore, — 61
Wilden forge, Worcs. 113
Wilkins, Jeffrey 19
  John 14, 19
  Walter 19
Wilkins & Jeffries 14
Wilkinson, Isaac 65, 116, 121, 127
  John 63, 65–6, 121
  William 66
Wilkinson's Executors 60
Willey Co. 63
Willey furnace, Shropshire 62–3
Williams, Henry 14, 33
  Phillip 15
Wilmott, James 113
  Robert 113
Wilson, John 116
Wilson House ?furnace, Lancs. 121
Wilton: *see* Eade & Wilton
Winchilsea, Earl of: *see* Finch, Heneage
Windsor, Herbert, 2nd Viscount Windsor etc 19, 22
Wingerworth furnace and forge, Derbys. 91, 95–7

Winter, Sir John 40, 47
Winter family 44
Wombridge furnace, Shropshire 63
Wood, Charles 21, 84
  William 72, 112, 125
Woolpitch Wood: *see* Trellech
Worcester, Earl of: *see* Somerset, Henry
Wortley forges, Yorks. 99–100, 102
Wrekin furnace, Shropshire 63
Wright, William 61
Wright & Prickett 135
Wriothesley, Henry, 1st Earl of Southampton etc 142

Yalden, John 138
Yarranton, Andrew 75
Yate, Appollonia 57
  Dame Mary 57
  Nourse 43
Ynyscedwyn furnace, Brecs. 33
Ynysygerwyn tinplace works, Glam. 44
Yonge, William 83
York Buildings Co. 145, 151
Younge, John Travers 102
Ystradgynlais, Brecs.: *see* Ynyscedwyn